坐月子調理良方

這書奉獻給各位產婦，
讓她們展開健康愉快的生活！

代序

「人的生命是靠能量來維持的，人體的能源主要來自食物，中醫理論認為：湯是食物中最富營養價值，亦很容易被人體吸收過來，食物中又分寒熱溫涼四氣及五味。」

在我多年臨床上，我一定囑咐所有孕婦必須安胎，所有食療方法是為着每個新生命做好基礎，令日後每一個小朋友都有強壯的體質，減少患病機會。本人所提供的食療方法十分簡單，在這可一同分享，方法如下：

孕婦每天宜吃：

鮮奶、雞蛋可多至 3 個、少吃多餐、必須吃新鮮食物，最好少調味品。

忌吃：

海產除了魚和蝦、肉類除了牛、豬和雞、生果宜吃蘋果和橙、菜類勿吃寒涼和帶有滑大腸作用的。

我在此建議所有孕婦最容易記的方法是，最適合吃新鮮而少調味料的食物，所有祛濕的湯水及食物都謝絕，坐月期間最好請教妳們所信任的中醫師。初懷孕至生產後整個過程最該相信的只有西醫或中醫。

能夠掌握眼前的食材，將之轉化為每個產婦身體必須的營養，是一個合格陪月員的責任，在這裏十分期盼每位讀者能夠通過這本專門食譜，達到融會貫通盡展所長的目的。

蔡潔儀校長在編著這本食譜時，經常與我討論研究如何能深入淺出地解說食譜的內容，從而令到每一位讀者都能夠容易明白怎樣運用各種食材的烹調方法，幫助每一位偉大的母親成功地渡過「安胎」和「坐月」的歲月。這樣的精神實在令人「肅然起敬」，在此希望這本精心傑作能暢銷風行。

國際中醫藥膳自療學會
創會會長註冊中醫師
盧壽如
弍零壹柒年春

2

孕婦的營養攝取和產後進補在中國人的社會非常重視，媽媽和奶奶都張羅着怎樣為自己的女兒或媳婦產後預備補品，有不同的進修課程或陪月班教人怎樣照顧產婦，我非常榮幸被蔡潔儀校長邀請為《坐月子調理良方》這本書作序。

在懷孕的早期，因為胎兒在發育階段，所以很多東西要特別注意，例如藥物，無論是中藥或西藥都要特別小心，如有疑問請咨詢你的醫生。

原則上食物要新鮮，含豐富的維他命、蛋白質、鈣、纖維質，避免醃製、太鹹、太甜、太肥膩的食物。

生產後因為很多女性都採用母乳餵養孩子，母乳是孩子最好的天然食糧，不但容易消化，而且增加孩子的抵抗力，但要顧及孩子的腸胃和有沒有敏感，如果食了母乳發生皮膚敏感，則母親要停止進食可能引起敏感的食物。

這本書為我們提供了豐富的內容，介紹了很多不同地方的習俗，適合產婦的菜譜，中醫湯水調理，甚至素食的朋友也可以在這裏找到適合自己的產後進補的方法，這是以前同類書少見的，讓很多人得益，值得一讀。

陳達明
腦神經外科醫生

序，是一個好開始！尤如新生命的降臨，為家庭帶來喜悅歡愉。

中國人傳統認為坐月子是媽媽的第二個春天，那是初為人母身體的新陳代謝重整期，若能好好把握調理，按個人體質進補得宜，那是可以令身體復元得更快更健康，並有足夠的營養乳汁餵哺小寶寶。傳統上，產後需要按不同階段調理近 3 個月，但現今都市人繁忙及化簡，一般坐月子則以一個月為限。

蔡校長在書中介紹精簡易明的坐月子調理資料及各式食材，在設計菜式上，以清淡為主，並附以產後 40 天的餐單，讓媽媽們吃得好、更易吸收營養。書中食譜葷素並重，切合現代人多菜少肉的飲食原則，亦顧及素食媽媽的需要，那是作者細心的巧妙心思也。

蔡校長見聞廣博，更介紹了湖北和台灣的坐月子補養資料，因地理環境的不同，補養方法與廣東有異，非常有趣及有耐讀性。

書中更推介了藥膳名方生化湯，那是中國傳統坐月的必服湯水，既有助媽媽，化生新血，服之可將惡露瘀血排出，新血自來之意。書中更介紹坐月必食的薑醋，其功能更有補虛、開胃、增乳、去瘀與增強子宮復元，那真是中國人的大智慧及平民恩物也。

蔡校長把多年在僱員再培訓局的教學心得結集成書，此書除準媽媽參考外，更適合現今的陪月員分享交流。

本人樂於為序，並推薦之！

徐欣榮（營哥）
國際中醫中藥總會會長

任教 ERB（Employees Retraining Board 僱員再培訓局）的烹飪課程好一段日子了，在課程中能為每個學員解決疑難，看着各人從一竅不通或幼嫩技能中成長，她們的努力，最後都能跨越萬難得到認同，是我最大的欣慰。

近兩年接任「陪月員小菜技巧班」，發現參與培訓的學員之中，不懂烹調或水平不高的佔大多數。因應學員們的需求，是為促使我編寫此書之原動力，產婦食譜不比尋常，有別於一般平常的烹調法，關係母子倆的健康問題，必須小心謹慎。菜式必須可提升產婦對飲食的興趣，也可平衡健康與美味之間的矛盾。

婦女生產後，坐月期間最為重要，能否恢復健康此為關鍵時刻，常聽長輩們説：「婚前體質弱的，產後坐月如調理進補得宜，不但健康能恢復，且更勝從前；反之便一落千丈，難以復元」，所以照顧產婦的飲食要特別小心，烹調法也以健康為主，少油、少鹽、少糖，少調味料，味精雞粉等都不應採用，但餚菜乏味又會使產婦欠缺食慾而吸收不到應有的營養等等問題，都是學員們惘悵的因由。正因如此，我將技巧竅門等等寫得清清楚楚；產婦的每日三餐，也編寫成多樣化可選擇形式的營養餐單表格，好讓產婦挑選自己喜愛的，陪月員也能清楚掌握，得心應手，使產婦和寶寶得到妥善照顧。

上課時，學生除了詢問一些與烹飪相關的問題外，也會問一些產婦遇到的日常問題，我亦盡量以我所知的回答，希望幫助到大家。但無論你可能是過來人又接受過專業培訓，從根本而言，再有經驗者，畢竟也只是陪月員而非專業醫生。我們必需知道，每個人的體質都不一樣，寒冷溫熱都不盡相同，A 婦食用的及照

顧方式若用於 B 婦可能未必合用，甚或適得其反。所以我建議學員們，產婦的健康、情緒、生活等等內在與外因的問題，若有疑難還是請教醫生為佳。

陪月學員勤奮好學，重情有禮，課餘之時敬我如親，視我如友，閒談間亦向我訴說她們工作的難處，和偶爾不被僱主所尊重之無奈。「陪月員」是為了產婦和初生嬰兒得到專業照顧而設的特別培訓課程，並非一般家庭傭工，應受到僱主的尊重。無奈，有很多僱主可能對此職位理解不足，而把她們視作一腳踢的女傭，家中大小事務，包括全屋清潔、買餸煮飯、照顧長幼等等與產婦和嬰兒無關的工作，都要陪月員處理。我在此特別強調此一誤點，倘若陪月員要分身如此多工序，便難以專心照顧產婦和嬰兒，質素及安全度必然下降，如此，對僱員及僱主都並無好處，相處亦難以愉快。倘若認為家傭與陪月的工作範疇是相等的，那政府又何需浪費資源去培訓一眾專業人才，希望聘請陪月員的僱主們留意這一點，不要加重陪月員的工作壓力，應建立互相尊重信任的良好關係。

此書內容，除了編寫適合產婦的飲食、解答了一些陪月學生的常見烹煮問題外，亦介紹了最具中西醫爭議的「生化湯」。此湯能排惡露、提高抗體、活血補虛兼可幫助子宮收縮，可說是先輩們必用的產後進補品，現今仍是台灣及內地婦產科常用的調養良方；惟是香港西醫及助產士，早在婦女產前檢查時已囑咐切忌服用。隨着社會變化，新舊概念交替，中西方醫術分歧，教人無所適從，如何選擇，便要看個人智慧！

　　在此，我特別鳴謝同心中醫治療院主席盧壽如中醫師和一本台灣出版物《坐月子的方法》。盧壽如中醫師為此書擔任顧問，給予多方面意見，並提供關於對婦女產後大有裨益的「生化湯」資料及服用方法，謹在此送上萬分謝意！《坐月子的方法》對我在懷孕期的幫助非常之大，尤其是手部按摩方法。當時的我身在遠方，親人在港，有不適時唯有閱讀友人送贈的書籍自療，並把部分資料和手部按摩方法記錄下來。今天有緣編寫此書，故將手部按摩方法與大家分享。

蔡潔儀

目錄

早餐

米飯

湯品

小菜

代茶

甜品

食療

*代表素食

*Vegetarian Recipe

懷孕概要——
寫在坐月子之前

寫有關坐月的食譜前，我想先分享一些民間智慧的安穩平和受胎法，懷上了，怎樣令胎兒和準媽媽身心健康，平平安安誕下麟兒。孩子出生時的喜悅，相信是每位父母永誌難忘的。

如何生個身心健康的好寶寶

每對夫婦都期待能孕育品性良好、身心健康的寶貝孩兒。

在中國古代，民間對受胎的時辰有明白的禁令，現在此略舉些較具體，而應注意的幾點供讀者參考，以下的時刻應避免性行為：

疲倦時、酒醉、天氣突變、生日、過度緊張、參加葬禮及吵架。

避過以上幾點，夫婦都有健康的身體和平順的心情讓孩子受胎，創造孕育優秀孩子的基本條件。

懷孕中的日常生活及家居安全

懷孕初期（100天內），最容易流產，要小心注意家居安全：

家中盡量減少濕滑的地方、避免容易摔倒、不要伸手取高處的東西、浴

室浴盆的邊沿和牆壁要有把手，以防滑倒。習慣每日記錄基礎體溫（在懷孕中，如基礎體溫突然下降，會有流產的可能，要立即請教醫生。），懷孕100天內避免性生活。外出時，陽光強烈要戴上太陽眼鏡。要盡量減少看書、看電視，不要讓眼睛疲倦。多聽柔和的音樂，抽烟飲酒更絕對不能。

懷孕期中有關的疾病

許多慢性疾病如扁桃腺炎、過敏性皮膚炎、腎炎、肝炎、胃炎、氣喘、風濕性關節炎、聽力障礙、肩胛酸痛、腦性麻痺多是來自感冒的後遺症。在懷孕期間，最容易患上感冒，但有很多藥物因對胎兒有影響都不可服用，如感冒特效藥、精神安定劑、賀爾蒙製劑等絕對不能吃。

因此，懷孕初期的婦女更要加倍小心照顧自己的身體健康，如多漱口、攝取充分的營養、盡量少到公共場所，在家安靜休息，因感冒最易發生在身體疲倦的時候。患病就要去請教醫生，不可隨便吃成藥。

懷孕期的飲食調理

懷孕初期（2-3個月）身子總是較怕冷，至懷孕3個月之後，體質轉為怕熱，因此懷孕期間的飲食需注重調理，在懷孕4個月之後可進行涼補。

在懷孕開始的6個月內要吃用綠豆、紅豆、青碗豆煲的骨湯、大魚頭湯等，因為母體這時需要鈣質，多飲牛奶、

懷孕期內宜享用的食物

蘿蔔

西蘭花、番茄、蘑菇都含有豐富的葉酸

多食有鈣的食物，至於鈣片之類的非天然補劑最好不吃。

　　海帶、蘿蔔、蓮藕（攪成汁喝）可預防便秘，有抑制脹氣的功效，多吃含葉酸的食物如西蘭花，補充葉酸越早越好。葉酸對早期胎兒腦部和脊髓的發育非常重要。

　　盡量避免吃辛香料、燒烤的食物、魚生、肝等，因為這些都是含有強烈興奮作用的食物，對胎兒有影響。此外，砂糖、鹽或醬油等混和的調味也不要吃，盡量吃單味的食物。

含葉酸的食物：

蔬菜： 菠菜、萵筍、番茄、胡蘿蔔、油麥菜、龍鬚菜、西蘭花、扁豆、
　　　　蘑菇等。

水果：橙、草莓、櫻桃、香蕉、檸檬、桃、李、杏、山楂、石榴、葡萄、奇異果等。

動物性食品：肝臟、肉及蛋類。

豆類、堅果類食品：黃豆、豆製品、核桃、腰果、榛子、杏仁。

穀物類：大麥、米糠、小麥胚芽、糙米等。

去除脹氣的按摩法

孕吐（一般稱作害喜），懷孕到了 3、4 個月時，有些孕婦會發生嘔吐的症狀，飲食的習慣亦因此而改變，這就是身體中存有脹氣的原因。通常孕吐容易發生在胎盤形成的階段，亦有些人到了 9 個月還會有孕吐的情形，以下是介紹一些預防孕吐的方法，照着做應該可以愉快地渡過這段非常時期。

飯前和睡前要做拉耳朵的按摩，也要做手部的按摩及肩胛骨的按摩。

手背的按摩法：

首先，左手掌的指頭要靠攏，平放於桌上，用右手的五隻手指按摩左手背，由手背到指尖，往手掌的方向用力的摩擦，直至手部溫暖後再換手。一般人認為噁心想吐之原因必然在胃部，其實不然，噁心可能是由手足發冷所引起。如發覺腳腕發冷，可以用溫度適合的熱水沖腳，以保暖的方法使之溫暖起來，加上手部的按摩，脹氣便會消除。

手的按摩法：

用一隻手為另一隻手做按摩，左右手相互按摩，由手腕、手背到指尖，按到指尖時，應稍用力刺激，因為刺激末稍神經可以促進胃部的蠕動，提高消化能力。

多攝取鈣質

懷孕 7 個月，快將進入孕晚期，腹部迅速增大，容易感到疲勞，有些孕婦還會出現腳腫、腿腫、靜脈曲張等狀況。所以每天一定要注意休息，避免經常彎腰，不要長久站立，晚上睡覺時把枕頭或坐墊墊在膝下，在飲食方面要多攝取鈣。

含鈣的食物：

乳類與乳製品（牛奶、奶粉、奶酪、酸奶），大豆與豆製品（黃豆、毛豆、青豆、黑豆、豆腐、豆干、腐乳），水產類（鯽魚、鯉魚、蝦、蝦皮、海帶、紫菜、海參），肉類與禽蛋（牛肉、雞肉、雞蛋、鴨蛋、皮蛋），水果類（檸檬、枇杷、蘋果、黑棗、山楂、杏脯、杏仁、合桃、瓜子、花生、蓮子等）。

懷孕 10 個月，孕婦不可一個人出遠門，因為隨時可能分娩，嚴禁性生活，這期間非常容易早產或受感染。每週去做一次產前檢查，堅持接受復查，要有足夠的睡眠和休息，儲備體力和精力，迎接健康好寶寶。

肉類、海帶、豆類、蛋等含豐富的鈣質

生產前的準備

預約陪月員

知道自己懷孕後，就要挑選陪月員。挑選陪月員，除了要看她們的證書和推薦信外，最重要的是要面談，這能互相了解對方的要求，你也可以感受陪月員是否有愛心、耐心和經驗。優秀的陪月員是十分「搶手」的，她們可以減輕媽媽們的負擔，可以全心全意休息，以補充元氣。

準媽媽

進入孕期最後一個月，孕婦會覺得時間漫長，心裏急着跟肚子裏的寶寶見面，情緒會變得焦躁，心神不寧。對於分娩，不少初懷孕的準媽媽感到恐懼，無所適從，甚至驚慌失措，其實生育過程是每位女性的本能，不用害怕，不必緊張。可以看多點相關資料，以了解生育的過程，做好心理準備。

準爸爸

準爸爸要多給妻子鼓勵、勇氣和幫助，陪伴妻子做運動和產檢，尤其是臨近生產前的產檢。為妻子做好一切準備，佈置好清潔舒適的房間，檢查寶寶的用品是否齊全，確認分娩時的聯系方式和交通工具的安排，一旦突然作動也不會太驚惶失措。

家人

家人可為產婦預先煲好薑醋，做好炒米，預早購買生化湯，如需要請陪月員的，更要預先與陪月員溝通，專業的陪月員是可以幫到你的。

甚麼是陪月員

　　所謂「坐月」，依照正常情況，順產要休息 30 天，不足月或剖腹生產則要 40 天。在這段期間，產婦一定要充分休息，而且要照顧得特別周到，所以陪月員便大派用場。專業的陪月員是經過專業培訓，她們一定可以在這段坐月期間幫到產婦，照顧她們一日三餐的合理營養飲食搭配，亦可幫手照顧嬰兒，讓產婦有充分時間休息使身體盡快康復。尤其是產後第一個星期的休息最重要，相信每一位陪月員都會懂得產婦食物的營養知識及衛生，和產後的飲食重點。所以請陪月員幫忙是個理想的選擇。

　　陪月員最好能為產婦準備五更飯（清晨 5 時進食），此時進食最能吸收。如用藥膳，產婦及其家人必須請教已註冊的中醫師，在其指導下，方可以藥膳進補，平常以清淡為主，烹調以蒸、燉、燜、煮為宜。產婦要少食多餐不能過量，每日以五至六餐最為適合，因產後胃腸功能減弱，少食多餐利於消化吸收。

　　特別在此一提的是，有某些僱主心存一種錯誤想法，以為陪月員在這段期間之工作，除了照顧產婦外，還需要為其家人服務，買菜煮飯等，甚至做一些不是為產婦而做的清潔工作。大家必須明白，陪月員是一門專業的工作，有別於一般家庭傭工，是特為產婦服務而受訓，她們的工作只照顧產婦和嬰兒，這點大家必須要留意及尊重。

食物營養

看我為各產婦撰寫的食譜之前，看看食材的營養價值吧！

紅糖的含鐵量比白糖高 1 至 3 倍，具有促進瘀血排出及子宮復元的功效，對於產後多虛多瘀的產婦，尤為適宜。

茄子（矮瓜）帶鹼性，不能生吃，生吃會有澀感。茄子主要含糖類、維他命、黃銅類化合物、脂肪、蛋白質、鈣、磷、紫蘇貳等。產婦常食茄子有活血化瘀、清熱消腫的作用，由於茄子中含有黃銅類化合物，具有抗氧化功能，可調節血壓及保護心臟。

紅糖

茄子

南瓜

南瓜含有豐富的維他命 A、B、C 及礦物質，以及人體必需的 8 種氨基酸和兒童必需的組氨酸、可溶性纖維、葉黃素和磷、鉀、鈣、鎂、鋅、矽等微量元素。南瓜還含有一種叫做「鈷」的成份，食用後有補血的作用。南瓜可以調整糖代謝，增強產婦的肌體免疫力，防止血管動脈硬化，同時有防癌的功效

冬瓜

每 100 克冬瓜肉中含蛋白質 0.4 克、碳水化合物 2.4 克、鈣 19 毫克、磷 12 毫克、鐵 0.3 毫克及多種維他命，特別是維他命 C 的含量較高。冬瓜幾乎不含脂肪、碳水化合物含量也很少，熱量低，屬於清淡食物，是含水量高的蔬菜。冬瓜能清熱解暑、利尿通便，有助於人體的清廢排毒，故特別適用於防治產後水腫、雙乳脹滿、痰喘、痔瘡等症。

每 100 克黑木耳含蛋白質 10.6 克、脂肪 0.2 克、碳水化合物 65.5 克、粗纖維 7 克，還含維他命 B1、B2、胡蘿蔔素、菸酸和無機鹽，鐵質的含量極高。由於黑木耳中鐵的含量極高，故產婦常吃可防治產後缺鐵性貧血。黑木耳還含有維他命 K，能減少血液凝結成塊，預防血栓等症。它的膠質更可把殘留在人體消化系統內的灰塵、雜質吸附集中起來，並最終排出體外，能起清理胃腸的排毒作用。

海帶

鯉魚

海帶是一種神奇食品，脂肪與熱量含量低，維生素亦微乎其微，但它卻含有豐富的礦物質（無機物），如鈣、鈉、鎂、鉀、碘、硫、鐵、鋅等以及硫胺素、核黃素、硒等，人體不可缺少的營養成分。

海帶含有大量的碘，因此可以刺激垂體，使女性體內雌激素水平降低，恢復產婦卵巢的正常機能，調理內分泌，消除乳腺增生的隱患。海帶富含硒，具有防癌的作用，常吃海帶對頭髮的生長、潤澤更具特殊的功效，能消腫去水和將體內毒素排出。

鯉魚對產婦有利尿消腫、益氣健脾、通脈下乳的功效。主治浮腫、乳汁不通等症；產後食慾不振者也適合食用。

鯉魚能促進乳汁分泌，如覺得豬腳太膩滯，不妨以燉鯉魚湯代替。

鯽魚

鯽魚的蛋白質含量為 17.1%，脂肪為 27%，並含有大量的鈣、磷、鐵等礦物質。坐月子喝鯽魚湯，是中國北方的古老傳統，自古以來鯽魚就是產婦的催乳補品，吃鯽魚可以令產婦乳汁充盈，鯽魚油有增強心血管功能、降低血液黏度、促進血液循環的作用。

鮑魚

鮑魚肉質鮮美，營養豐富，蛋白質含量高，鮮品中含蛋白質 20%，乾品高達 40%，含人體必需的 8 種氨基酸，是海產品中的珍品，與海參齊名。產婦吃鮑魚不單可以清肝明目、防治高血壓、腫痛等症狀，還可以增乳、益智。

黃鱔

黃鱔性溫、味甘，具有補氣養血、補肝健脾的功效，還有通乳、增乳的作用。此外黃鱔的頭骨內服可止痢，鱔皮可用於治療乳房腫痛。

金針

青椒

金針即金針菜，簡稱「金菜」或「針菜」，是黃花菜的曬乾花蕾，別名「忘憂草」。含優質蛋白質，可提供人體必需的 18 種氨基酸，其中以精氨酸、賴氨酸含量最豐富，能清熱解毒、止渴生津、利尿通乳，又可治口乾舌燥、大便帶血、小便不暢及便秘等症狀。注意：剛採摘的鮮金針含有秋水仙鹼，帶有一定毒性，如果大量食用，容易引起中毒，所以宜曬乾後再食用。

青椒又叫「甜椒」、「柿子椒」、「西椒」，維他命 C 含量高，每 100 克青椒中含碳水化合物約 3.8 克，一個青椒雖然只有 35 卡的熱量，卻能供應多種維他命及礦物質。除對產後補血有好處外，還有減輕妊娠斑的功效。

百合營養豐富，含蛋白質、蔗糖、還原糖、果膠、澱粉、脂肪、生物素、水仙鹼、磷、鈣、維他命 B2 等營養成分。百合高鉀低鈉，能預防產後高血壓，並可保護產婦血管的作用，更有效改善產婦貧血和排毒，尤其適宜產後抑鬱症者食用。

書內以鮮百合入饌，它的味道清甜，容易入口，定必能引起產婦的食慾。選購百合時，以表皮潔白，沒有累累傷痕、乾爽、瓣塊緊密包覆的為佳。

通草為五加科灌木植物通脫木的莖髓，於秋季割取莖枝，它含有肌醇、多戊糖、多聚甲基戊糖、阿拉伯糖、乳糖、半乳醣醛酸等成分。能利小便、下乳、明目、退熱、催生、治目昏耳聾、鼻塞失音，可治水腫、產婦乳汁不通。

選購通草時，以顏色潔白，心空、有彈性者為佳。它的味道很淡，容易入口，在這書，它配以豬腳煲湯，也用於代茶。

產後飲食之道

產後的調養食物，* 有些地方以麻油炒老薑為主料（*詳細資料看 P.27），再把煮好的材料如豬腰、豬肝、雞或糯米加入一起烹調。每一道食品都先以老薑炒香至褐色，作用是刺激體內臟器，使其活潑化，讓身體內部暖和起來。此外在烹調產後調養品時，米酒是不可缺少的，因為加了它，可改善產婦的血液循環，而且能把懷孕期中積存的廢物，完全地從體內排出。

有人說，開刀的產婦不能飲酒，其實無妨。採用自然生產的產婦，接生的醫生有時也會剪開產婦的會陰，以利生產，同樣會有傷口，只要傷口不發炎化膿，加些酒來作烹調是沒有壞處的；況且飲食中所加的酒，在烹調過程中，其酒精成分早已揮發得差不多了。

不要聽到脂肪兩字就害怕，其實脂肪在人體營養中佔重要地位，產後的飲食中所提供給產婦的脂肪，對乳汁的分泌和乳汁中的脂肪成分有密切關係。產婦若脂肪攝取不足，便要動用身體內儲備的脂肪，長此下去，對嬰兒和產婦本身都沒有好處。因為產婦體內的脂肪酸有增加乳汁分泌的作用，而嬰兒的發育及對脂溶性維他命的吸收也需要足夠的脂肪。因此，在膳食中必須有適量的脂肪才可供給自己和嬰兒的身體所需。

進行母乳餵哺的產婦，體內的鈣會隨着乳汁大量消耗，如果沒有及時補充鈣，就會引起腰酸背痛、腿腳抽筋、牙齒鬆動、骨質疏鬆等，還會影響嬰兒身體各項發育，因此新任媽媽每日應多吃豆腐、雞蛋、魚和牛奶等，這些食物除含豐富的蛋白質外，還含有大量的鈣。

產婦應如何進補

產後 40 天內可以進補嗎？

產後 1-2 週，產婦要排出體內的惡露，同時亦要促進順產時陰道撕裂或剖腹生產後的傷口愈合。在這 1-2 週的關鍵期裏，產婦不宜大補，因在這段時期，大補很容易導致血管擴張加劇出血，而延長子宮的恢復期，造成惡露難以排清，因此，在產後頭 1-2 週內，應給產婦清淡及稀軟的食物。如是順產者，可以開始進食薑醋。

第 15 至 24 天可以開始溫補，以助消化、排便機能恢復，補腦固腰腎，但不宜太燥。產後第 25 至 31 天要開始食用養血、補元氣食物。產後第 32 至 40 天宜食用健脾胃、滋陰潤肺、恢復體力的食物。

產婦 40 天過後還要進補嗎？

生產過了 40 天後，是可以逐漸加入某些藥材和食物並用，經烹調精製成美味的養生佳肴，以藥膳來補身。

產婦進補最理想是補足 100 日，採用中藥來養生，是我們的老祖宗留存下來所累積的智慧。藥膳是中國人特有的一種烹調技術，是以中藥材和食物並用，而增加了藥材的滋補調養功效。

如以藥膳進補，又應該吃些甚麼呢？

人參、石斛、鹿茸、鹿尾巴等都是產婦所需要的補藥。藥食同源，經過了中醫學專家們的諸多實踐，累積經驗，讓我們了解到它特有的四氣（寒、涼、溫、熱），五味（酸、苦、甘、辛、鹹），歸經（治療的部位）。

要請教中醫嗎？

由於每個人的體質有別，故不同體質便有不同的補法，並不是所有產婦都可以食同一種藥膳來進補。40 天過後，我建議先去請教中醫師，把把脈，瞭解清楚自己的體質及四氣（寒、涼、溫、熱），不同人身體出現不同情況，補法亦有不同。

例如：年已過 32 歲的產婦，中醫師會建議加入鹿尾巴來作藥膳，產後頭風重的，可加入天麻、女貞子等，但加多少？怎樣搭配才是適合和最理想？這必定要請教中醫師，因每個人的體質有別，不能任意搭配。

產假過後，怎樣休息和補養？

一般產假約 6-8 週，產婦身體還未完全康復便要上班，這真是不簡單的事，因此，她的家人必須懂得諒解和協調。全日的工作，到了下午 3、4 點左右最易疲倦，所以下班回到家裏，宜先休息一下，待精神稍為恢復再做家務。在晚上先把明天的早餐和午餐準備好，由於還未過 100 天，所以最好自備飯盒，做些清淡的食物，盡量避免在外食飯，到了週日放假的時間，盡量燉些補品來填腹。

初生蛋

其他地方的產後補養方法

湖北

湖北廣水，是我大媳婦的家鄉。

她們在坐月期間的習俗，是每天都要吃6至7個土雞所生的「初生雞蛋」。

初生蛋的個子非常細小，據她們說，初生雞蛋是最佳的補品。早餐吃最好（可一次過吃完），做法是以紅糖水煮雞蛋，還要加一塊肥豬肉（又稱白肉），約40克同煮。她們每次煮甜食，都會放一塊肥肉（她們稱之為豬油），據說這樣能使皮膚潤滑。

以下是當地的催乳秘方，據說是十分有效。

糯米、紅棗、花生加入已切成丁粒的肥豬肉，與水一起煮至糯米糜爛，然後加入紅糖或冰糖，再煮至糖溶，便可食用。

至於常用的補身湯水，有鯽魚煲蘿蔔、鯽魚豆腐湯、蓮藕排骨湯、花生煲豬手湯等，這些湯水除補身外，還有催乳的作用。

甜食有酒釀煮雞蛋、紅棗銀耳花生煮紅糖或冰糖。

主食方面，則多以麵條及餃子為主。

至於產後習俗方面，婦女產後不吃薑醋，她們認為吃酸醋會使鈣質流失，令骨質疏鬆。產後只可用水漱口，一星期後才可以擦牙，否則牙齒容易鬆脫。坐月期間第一個月不能沖涼，只可用熱水抹身，同時要待滿月才能洗頭。

台灣

產婦在分娩後，就立即以＊生化湯來填腹，她們習慣不論是順產或剖腹生產，在嬰兒出生後一週內，每天都要飲用生化湯，因為生化湯以養血活血為主，而且更能排清惡露及幫助子宮收縮，她們認為如產後生化湯和麻油雞酒喝得不足，就會產後嚴重便秘、發燒、感冒，並會有子宮下垂的現象。

據所知，生化湯裏的當歸和桃仁有通便作用，而麻油雞酒又利於清除體內積存了 10 個月的廢物，使子宮的收縮能力復元，促進新陳代謝。

產後第一週要吃連皮老薑，以黑芝麻油炒成褐色，加入豬肝，用大火快炒，再加入米酒 1/4 杯（約 60 克）煮滾，立即熄火，湯汁趁熱喝，老薑和豬肝用來做下飯菜。第二週以豬腰取代豬肝，做法與豬肝相同，用黑糖、老薑或米酒煮黑豆或紅豆吃。第三週以麻油雞酒代替第一、二週所吃的豬肝和豬腰。

調味方面：鹽在產後 30 天內禁止使用。酸味、鹹味都不能食用，因為酸味會使內臟下垂，鹹味會增加腎臟的負擔和骨頭的軟化，食物越淡越好，以原味為最佳。

＊註：根據古書記載，「生化湯」以養血活血為主，所以普遍用於婦女產後補血、祛惡露。此湯不但可以活血補虛，更可以提高抗體力量，對子宮有明顯的收縮作用。

黑芝麻油

老薑

上奶的補充：產婦生產後第三天開始適量的吃花生香菇老薑活蝦煮豬腳，做法如下可作參考：

材料：香菇、花生、老薑、豬腳、黑芝麻油及活蝦

製法：

香菇洗淨搾乾水，浸在過面的米酒中過夜，切絲待用。花生仁洗淨；豬腳斬件飛水瀝乾。老薑洗淨連皮切片。

麻油起鍋，炒至老薑淺褐色，加入豬腳炒至上色，然後加入花生及香菇炒透，最後下活蝦和米酒，加蓋燒滾後，改以小火煮３小時即成。用來佐膳，但盡量於早、中餐時食用。

這菜式可做多些放入冰箱保存，待吃時加熱食用。還有黑芝麻油雞酒、油飯（糯米炒飯）、豬腳麵線等，都是台灣婦女在坐月時的珍品。

此外，和廣東習俗相若，她們要待滿月後才可以洗頭。

產後第一週餐單

開刀的產婦除了薑米麻油炒飯外，其他皆可進食，因這食譜以薑為主料怕影響傷口。其餘食譜的薑作料頭用，份量少，取其增香辟腥之效，對傷口的影響輕微甚至沒有。

早餐	五更飯、紅糖粥、桂圓杞子紅棗粥、紅棗南瓜肉碎粥、雞火煨麵 建議早餐（不設食譜）：焓熟雞蛋
米飯	鯽魚上湯魚片蒸飯、蝦皮薑米麻油雞蛋炒飯（素食者可刪去蝦皮和雞蛋）
湯品	清燉雞汁或清燉肉汁、紅棗木瓜花生湯、佛手南瓜番茄冬菇湯、蓮藕章魚豬手湯、海帶排骨湯
小菜	可以白飯或白粥（白粥要加陳皮煲）配以下小菜： 清煮南瓜、豉油雞 建議小菜（不設食譜）： 梅菜蒸肉餅、清蒸魚、冬菇紅棗蒸滑雞、西蘭花番茄豆腐、南瓜煮肉碎
蔬菜	菠菜、西蘭花、椰菜、莧菜、菜心、芥蘭、青甜椒、紅甜椒、蜜豆、粟米、節瓜、南瓜、番茄、薯仔、紅蘿蔔、茄子、冬菇、秀珍菇、草菇 以上蔬菜均可在坐月期間食用，以薑片或蒜頭起鑊，烹調法炒、煮、浸、灼均可。
代茶	炒米茶、通草北芪茶、紅棗炒米茶
甜品	紅棗合桃露、杏汁鮮奶燉木瓜 建議甜品（不設食譜）：蓮子鮮百合糖水、芝麻糊
水果	木瓜、提子、西梅、草莓、蘋果、橙、哈密瓜 以上水果，產後可食用。宜禁荔枝和榴槤，尤其需要哺乳的媽媽、
零食	乾無花果、合桃、松子、杏脯、南棗合桃糕、桂圓合桃糕、番薯乾

註：
- 建議的五個餐單，有些菜式的做法太簡單，故沒有提供食譜。
- 綠色的是素食
- 清晨 5 點時吃的飯，稱為五更飯，有容易吸收及補氣之效。

產後第 8 至 14 天餐單

產後第二週了，可以吃第一週和今個餐單，除了薑醋外，開刀生的產婦其餘也可食用。

早餐	紅糖粥、桂圓杞子紅棗粥、瑤柱花膠雞肉粥、雞火煨麵、焓雞蛋、五更飯（於清晨 5 時吃飯）
米飯	鯽魚上湯魚片蒸飯、栗子飯 建議米飯（不設食譜）：梅子排骨飯
湯品	蓮子菠菜銀耳湯、紅棗木瓜花生湯、清燉雞汁或肉汁、 通草豬腳湯、花膠椰肉燉烏雞、栗子雪耳鯽魚湯 建議湯品（不設食譜）： 栗子雪耳紅蘿蔔湯
小菜	冬菇扒豆苗、海帶滷冬菇、原條鮮魚蒸滑蛋、雞蛋豬腳薑醋 建議菜式（不設食譜）： 青椒木耳炒肉絲、梅菜蒸豬肉、冬菇豆干煮節瓜
蔬菜	參照產後第一週餐單
代茶	參照產後第一週餐單
甜品	薑汁蛋白燉鮮奶、鮮百合南瓜露 建議甜品（不設食譜）： 陳皮紅豆沙、木瓜雪耳糖水
水果	參照產後第一週餐單
零食	參照產後第一週餐單

產後第 15 至 24 天餐單

開刀生的產婦，除了魚頭雲煮酒、雞蛋豬腳薑醋，其他食譜均可食用

早餐	五更飯 建議早餐（不設食譜）： 桂圓蓮子粥、柴魚花生豬骨粥、鮑魚雞絲通粉
米飯	栗子飯、黃鱔蒸飯、補血豬膶飯、金針雲耳田雞煲仔飯、蝦皮薑米麻油雞蛋炒飯
湯品	鮑魚北芪燉竹絲雞、栗子雪耳鯽魚湯、補血豬膶水、番茄魚茸羹、雙棗黑豆燉鯉魚 建議湯品（不設食譜）： 紅棗桂圓蓮藕湯、木瓜花生無花果雪耳湯、黑豆紅棗煲塘虱、杞子淮山烏雞湯
小菜	羅漢滑豆腐、菠菜木耳炒香干、肉桂蜂蜜醬油雞、鮮百合蜜豆炒肉片、魚頭雲煮酒、雞蛋豬腳薑醋 建議菜譜（不設食譜）： 清蒸魚、蒸滑蛋、瑤柱冬菇蒸肉餅
蔬菜	參照產後第一週餐單
代茶	參照產後第一週餐單
甜品	桂圓杏仁茶、腰果露、江南桂花酒釀丸子、鮮百合南瓜露。 建議甜品（不設食譜）： 芝麻糊、薑汁撞奶
水果	參照產後第一週餐單
零食	參照產後第一週餐單

產後第 25 至 31 天餐單

除了麻油老薑雞煮酒、雞蛋豬腳薑醋外，以下所有食譜開刀生的產婦均適合食用。

早餐	紅棗南瓜肉碎粥、桂圓杞子紅棗粥、五更飯 建議早餐（不設食譜）： 栗子百合番薯粥、蠔豉瘦肉粥
米飯	栗子飯、日式蜆肉飯、補血豬膶飯、鯽魚上湯魚片蒸飯 建議米飯（不設食譜）： 梅子排骨飯
湯品	佛手南瓜番茄冬菇湯、蓮子菠菜銀耳湯、**雙棗黑豆燉鯉魚**、花膠椰肉燉烏雞 建議湯品（不設食譜）： 猴頭菇花膠煲豬腒、栗子合桃花膠煲雞胸肉、肉茸瑤柱紫菜羹
小菜	糖醋排骨、雞蛋豬腳薑醋、麻油老薑雞煮酒、冬菇紅棗蒸斑片、香煎蜜汁銀鱈魚、原條鮮魚蒸滑蛋、美果炒雙珍、香芹炒藕片、海帶滷冬菇 建議小菜（不設食譜）： 茄子煮魚鬆、栗子燜雞
蔬菜	參照產後第一週餐單
代茶	參照產後第一週餐單
甜品	桂花糖釀蓮藕、腰果露、紅棗合桃露、**桑寄生蛋茶** 建議湯品（不設食譜）： 冰花燉蛋
水果	參照產後第一週餐單
零食	參照產後第一週餐單

產後第 32 至 40 天餐單

這裏的全部菜式，不論是順產或開刀的產婦都可以吃，不用禁食。

早餐	紅棗南瓜肉碎粥、瑤柱花膠雞肉粥、木耳菠菜肉碎餃子、五更飯 各式不油膩包點（不設食譜）
米飯	日式蜆肉飯、黃鱔蒸飯、金針雲耳田雞煲仔飯
湯品	龍眼牛䐒湯、海帶排骨湯、栗子雪耳鯽魚湯、蓮子菠菜銀耳湯、 蓮藕章魚豬手湯
小菜	鮮百合蜜豆炒肉片、麻油老薑雞煮酒、魚頭雲煮酒、冬菇紅棗蒸斑片、糖醋排骨、肉桂蜂蜜醬油雞、菠菜木耳炒香干、香芹炒藕片、美果炒雙珍、冬菇扒豆苗 建議小菜（不設食譜）： 陳皮牛尾、栗子燜雞
蔬菜	參照產後第一週餐單
代茶	參照產後第一週餐單
甜品	桑寄生蛋茶、紅棗合桃露、桂圓杏仁茶 建議甜品（不設食譜）： 花膠蓮子百合燉鮮奶
水果	參照產後第一週餐單
零食	參照產後第一週餐單

在產後的一個多月期間，產婦要遵守食物的調養法，要好好把握這個補身的機會，若是錯失時機，對婦女而言，身體就有天壤之別，希望你們產後都能過着健康而調和、美滿的生活。

Steamed pork dumplings with
spinach and wood ear fungus P.158

木耳菠菜肉碎餃子

材料

木耳（浸透）.................. 20 克
菠菜 .... 150 克
攪碎豬肉 .... 150 克
薑茸 .... 10 克
餃子皮（白皮）............ 150 克

醃料

蠔油 .... 1 湯匙
醬油 .... 2 茶匙
鹽 .... 1/4 茶匙
生粉 .... 2 湯匙
麻油 .... 2 茶匙

做法

1. 木耳浸透，洗淨剁碎。

2. 菠菜洗淨，略飛水，過冷搾乾，切至極碎。

3. 碎肉置盤中，先將醃料加入攪勻，再把以上材料和薑茸加進再攪透。

4. 以餃子皮包成水餃，蒸碟塗些油，放上餃子，入鍋蒸熟約 10 分鐘。

烹調竅門

用來煮或蒸的餃子皮有甚麼分別？

一般以方形大白皮用來做雲吞，以水煮熟；圓形皮則做餃子，蒸或煎均可。另包剩之肉餡，可當肉餅蒸熟享用。

紅棗南瓜肉碎粥

Ground pork congee with
red dates and pumpkin P.158

材料

南瓜 ....300 克
攪碎豬肉 ....100 克
白米 ....120 克
清水 ....10 杯或適量
紅棗（去核、切片）.........4 粒

醃肉料

醬油 ....1 茶匙
生粉 ....半茶匙
熟油 ....1 茶匙

調味料

鹽 ....適量

做法

1. 南瓜去皮，切丁方粒；肉碎加醃料拌勻。

2. 白米洗淨置煲中，注入清水待滾，改用中火煮至白米開花，加入紅棗、南瓜再煮至粥綿，加入碎肉拌勻，續煮 10 分鐘即成，可加鹽調味。

註：此粥較濃稠，宜產婦享用，可代替米飯。

烹調竅門

怎樣才能令粥快點綿和不黏鍋？

最好備有高身的瓦煲，蓋頂上有小孔的，這是正宗的煲粥鍋具。用瓦煲煲粥，粥滑而綿，煲粥時，最好不要多手攪動，因攪動會黏底，待粥煲好時，才攪勻上鍋。

材料

瑤柱 .... 1 粒
嘉美雞 .... 半隻
濕發花膠 .... 100 克
薑絲 .... 10 克
白米 .... 150 克
雞湯 .... 10 杯

調味料

鹽 .... 適量

做法

1. 瑤柱洗淨；花膠切粒。

2. 雞洗淨飛水置煲中，注入清水熬成雞湯約 10 杯。

3. 起出雞肉切碎，雞湯隔渣備用。

4. 白米洗淨置煲中，注入雞湯、瑤柱，待滾後改用
   中慢火煲至白米開花，加入花膠續煲至粥綿。

5. 再加入雞肉碎及薑絲煮片刻，可下鹽調味享用。

烹調竅門

需要剝去雞皮才熬湯嗎？

熬雞湯、煲雞粥或用作上湯，如保留雞皮，湯雖然香
濃可口，但由於雞隻的脂肪多積聚於皮下（即黃油
雞），多食無益，所以需要剝去。

但如用的是嘉美雞，則雞皮是可以保留的，因嘉美雞
無論是雞項（雌性）或是嘉美少爺雞（雄性），皆是
皮薄及皮下脂肪少（是白皮的）。

瑤柱花膠雞肉粥

Chicken congee with
dried scallops and fish maw P.159

雞火煨麵

Soup noodles with chicken and ham P.159

材料

乾葱（切片）............ 1 粒
嘉美雞...................... 半隻
金華火腿................. 40 克
鮮雞上湯.................. 4 杯
陽春麵.................... 200 克

調味料

鹽...................... 適量

做法

1. 熱鑊以適量油炒香乾葱。

2. 雞洗淨，起出雞胸肉，蒸熟（撕成雞絲），餘下之雞件飛水過冷，瀝乾備用。

3. 火腿蒸熟（約 8 分鐘），取出切片（取 2 片切絲），餘下留待煮湯用。

4. 鍋中注入清水約 8 杯，加入已飛水之雞件及餘下之火腿，待沸滾後，改以中慢火熬成 4 杯雞火上湯備用。

5. 清水煮滾，加入陽春麵，煮至麵心熟透取出，過冷瀝乾。

6. 鮮雞上湯置瓦鍋中，加鹽調味，放進以上麵條，以大火煨麵 5 分鐘即成，把爆過之乾葱、雞絲及火腿絲放面上即成。

烹調竅門

甚麼是陽春麵？在那裏購買？可否用蛋麵代替？

陽春麵即相等於我們廣東人叫的淨麵。麵白而幼，入口嫩滑，在超市有乾貨出售，非常普遍。

如用蛋麵做這款主食，則味形口感都會失真不好吃。

紅糖粥

Congee with dried tangerine
peer and cane sugar

P.160

材料

白米	120 克
片糖（紅糖）	1 磚
陳皮（浸透）	1 小塊
清水	8 杯

做法

1. 白米洗淨放煲內，加入陳皮，注入清水。
2. 滾後改以中火煲至白米糜爛，加入片糖，續煮至粥綿即成。

烹調竅門

市售的紅糖質素參差，顏色有深有淺，那款才合用？

市售紅糖一般有塊狀黃糖（片糖）及黃砂糖兩種。片糖的提煉程序較砂糖少，而且較濕，保存之香味較多，維他命及礦物質亦較多量，但也因此含雜質較多，用時最好沖洗表面，或以小量水溶煮成水漿後，棄去聚在底處之沉澱物質。選用紅糖，一般以顏色淺淡為佳。

Congee with dried longans,
goji berries and red dates P.160

桂圓杞子紅棗粥

材料

桂圓（乾龍眼肉）	15 克
杞子	1 湯匙
紅棗（去核）	6 粒
白米	120 克
清水	8 杯

早餐 素

做法

1. 白米洗淨；桂圓、紅棗與杞子分別略洗淨。

2. 除杞子外，將所有材料同置煲中，滾後以中慢火煲至白米糜爛。

3. 加入杞子再煮 15 分鐘即成。

烹調竅門

為甚麼杞子要最後才放入？

杞子煮前要清洗，洗去硫磺等防腐物質。杞子不宜久煮，因味道會變酸，不好吃。

註：這粥功效養心安神、健脾補血。主治心脾兩虛、頭暈心悸、健忘失眠。

金針雲耳
田雞煲仔飯

Clay pot rice with frog,
day lily flowers and cloud ear fungus P.161

材料

田雞	2 隻
金針	10 克
雲耳	10 克
紅棗（去核、切片）	3 粒
薑（切絲）	2 片
白米	240 克
清水	280 克

醃料

醬油	1 茶匙
蠔油	1 湯匙
鹽	1/4 茶匙
薑汁酒	1 湯匙
生粉	2 茶匙
熟油	1 湯匙

做法

1. 田雞處理乾淨後斬件，以醃料拌勻。

2. 金針、雲耳分別浸透，洗淨，搾乾水。金針打結。

3. 將以上材料同置盤中，加入薑絲及紅棗拌勻備用。

4. 白米洗淨瀝乾水，置瓦煲中，注入清水，以慢火煮約 15 分鐘。

5. 見飯面水分快將收乾，將田雞料平均鋪在飯面，再以慢火煮 15 分鐘，熄火再焗 5 分鐘即成。

烹調竅門

如不是用瓦煲煮，可以用電飯煲代替嗎？味道有很大的分別嗎？

如沒有瓦煲，可改用蒸飯的方法去做；如用電飯煲，質感、食味都減半，但你可將餸與飯分開做，用電飯煲煮飯，餸可另外蒸。

用明火煮的瓦煲飯，實在非常美味，雖然會花點時間，但是值得的。

註：田雞能大補元氣，治脾虛，適合精力不足，尤宜是胃弱或胃酸過多的產婦。

黃鱔蒸飯

Steamed rice with yellow eel P.162

材料

黃鱔	320 克
薑（切絲）	2 片
葱（切絲）	2 條
白米	240 克
清水	260 克

調味料

薑汁酒	1 湯匙
醬油	1 湯匙
蠔油	1 湯匙
麻油	少許
胡椒粉	少許
鹽	1/4 茶匙
生粉	1 茶匙
熟油	1 茶匙

做法

1. 黃鱔洗淨，用熱水略拖，洗去黏液，切粗條備用。

2. 鱔條置碗中，加入調味及薑絲拌勻。

3. 白米洗淨，瀝乾放蒸鍋中，注入清水，隔水蒸至飯熟約 20 分鐘，再把醃好之黃鱔平均放飯面，冚蓋再蒸 8 分鐘即成，食時可放葱絲及醬油。

烹調竅門

如用電飯煲代替隔水蒸飯，有甚麼需要注意？

如用電飯煲煮飯而飯面上又要放上其他生料，最好是剛跳掣便將生料平均放在飯面，再將煮飯的掣按下，如是者重複多按幾次直至生料熟透。但我個人覺得並不好吃。

蒸飯如蒸餸一樣易做，並不難處理，而且蒸飯有保身之功效。

註：黃鱔蒸飯能補血氣、壯筋骨、祛風濕、填虛損。

材料

豬肝（豬膶）.............160 克
白米 ....240 克
清水 ....280 克
薑絲 ....2 片份量

調味料

薑汁酒.............1 湯匙
醬油 ....1 茶匙
蠔油 ....1 茶匙
熟油 ....1 湯匙
生粉 ....1 茶匙
麻油 ....1 茶匙
胡椒粉.............少許

做法

1. 豬肝洗淨切片，用調味撈勻備用。

2. 白米洗淨放飯煲中，加入清水煲成飯。

3. 待飯即將收乾水分時，加入豬肝和薑絲放飯面上，焗至飯熟即成。

烹調竅門

處理豬肝有甚麼竅門呢？

豬肝口感嫩滑人人愛，要煮得美味，最重要是懂得控制火候及烹調法。豬肝切好後，用清水洗去血污，然後用少許生粉及調味料拌勻，待飯面快乾水時便可放進豬肝，待飯熟透，豬肝亦剛好恰到好處。如用作炒，工序會較多，豬肝切好後以清水洗去血污，加入醃料拌勻，要飛水至七成熟，再以清水將它洗去所附的血污及醃料（因這樣亦不至稍被風吹即變黑色），然後再洗淨鑊，加油爆料頭及配菜，把豬肝回鑊，灒酒及加調味打芡上碟。

注意：炒豬肝的動作要快，否則易變老不好吃。

補血
豬膶飯
Pork liver rice P.164

鯽魚上湯
魚片蒸飯

Steamed rice with
sliced fish and fish stock P.161

材料

鯽魚	1 條
陳皮（浸透切絲）	1 湯匙
白米	240 克
鯽魚湯	260 克

醃料

熟油	1 湯匙
生粉	1 茶匙
鹽	少許

調味料

鹽	1 茶匙

做法

1. 鯽魚處理乾淨，起肉切片。
2. 魚肉置碗中，加入醃料及陳皮絲拌勻。
3. 魚頭、魚骨、腩用油煎香，注入適量清水熬成魚湯 260 克，加鹽 1 茶匙調味，拌勻。
4. 白米洗淨瀝去水分置深碗中，注入魚湯，轉放蒸鍋上，以大火沸滾水蒸約 20 分鐘至飯熟。
5. 將已醃好之魚片平均放在白飯上，再蒸 5 分鐘即成。

烹調竅門

將鯽魚起肉切片有甚麼竅門？

在市場買回來的活魚，一般魚販已代宰殺，回家後洗淨抹乾，再進行如下的步驟：

切去頭部、尾部，把魚頭斬開兩邊（用來煎香後再熬湯），斜刀片去魚肉上的胸骨，沿着背骨片出兩片魚肉，然後切片就成。

註：

鯽魚性溫無毒，富含蛋白質、脂肪、鈣、鐵、磷等，營養價值甚高。它除濕利水，對脾胃虛弱、水腫、支氣管炎、哮喘、糖尿病等症，有極好之食療滋補作用。中國各地民間常以鯽魚製湯給產婦，以助通乳下乳。

日式蜆肉飯

Japanese clam rice P.163

材料

清水	1 千克
昆布	1 小塊（10 克）
蜆肉	200 克
裙帶菜（切碎）	10 克
三色米（糙米、紅米、白米）	600 克

調味料

醬油	2 1/2 湯匙
鹽	1 茶匙
味醂	2 湯匙
白酒	3 湯匙
薑汁	1 湯匙

做法

1. 先將清水和昆布放鍋中煮沸,再放入蜆肉,待沸滾後盛起,湯汁留用(昆布棄去)。

2. 調味和蜆肉拌勻,醃漬 30 分鐘後,將蜆肉撈出,瀝乾備用。

3. 將醃過蜆肉之調味加入湯汁中拌勻,然後過濾,留起 675 毫升上湯作煮飯用,剩餘之湯汁用作浸漬裙帶菜(使其入味)。

4. 待裙帶菜浸透後,便可盛起擠乾。

5. 白米洗淨,重複搓洗 2-3 次,瀝乾水分,靜置 30 分鐘後,加入上湯煮熟成飯。

6. 飯煮好後,加入蜆肉和裙帶菜,再焗片刻(約 10 分鐘),便可拌勻進食。

烹調竅門

在那裏購買蜆肉?怎樣挑選優質蜆肉?

市面出售的蜆多用水浸着,活蜆水清而殼緊閉,或蜆殼有開合動作,並冒出氣泡,都算是新鮮。如蜆殼已部份敞開不動,露出蜆肉,養蜆的水出現渾濁,均不宜購買。除新鮮活蜆外,亦可選購急凍或曬乾的,前者可在超市購得,曬乾的可到海味乾貨店購買,靚貨是沒有沙粒的。

註:

不論鹹水蜆或淡水蜆,所含的營養成分都大至相同。

將新鮮的淡水蜆肉煮酒飲,能增加乳汁分泌,並能促進產後精神幫助體力復元。

如用作食療,可將蜆連殼一同煮湯飲用,因蜆殼在藥用上有制酸作用。而且原隻生蜆裏面所含「蜆水」有清熱、去積、利咽喉、去痰涎的功用。

蝦皮薑米麻油雞蛋炒飯

Egg fried rice with dried shrimps, ginger and sesame oil P.162

材料

蝦皮	20 克
薑	20 克
雞蛋	2 個
白飯	2 碗
麻油	2 湯匙

調味料

鹽	適量

做法

1. 蝦皮洗淨瀝乾；薑去皮洗淨，切碎如米粒狀；雞蛋打散。

2. 熱鑊下麻油 2 湯匙，爆香薑米、蝦皮。

3. 蛋淋在白飯面，一起傾落鑊中，以大火炒散飯粒，加入少許鹽調味即成。

烹調竅門

甚麼是蝦皮？

蝦皮是指經曬乾的蝦幼苗，含豐富的鈣質。炒飯前，用薑米爆炒，除可辟腥外，更可增香。

註：

蝦皮味甘，性溫，鈣、鎂質豐富，對身體虛弱以及病後調養的人，有極好的食補作用，適合中老年、孕婦、產婦及缺鈣至小腿抽筋者食用。

栗子飯

Chestnut rice P.163

材料

白米 ....240 克
煮栗子水 ....280 克
鮮栗子肉 ....18 粒
清水（煮栗子用）............300 克

做法

1. 栗子肉洗淨放煲內，注入過面清水
   （約 300 克），以中慢火煮 20 分
   鐘至水分剩餘 280 克，將栗子盛
   起，汁水留用。

2. 白米洗淨，瀝去水分，放電飯煲
   內，加入焓栗子肉之汁水，及煮過
   之栗子肉，按掣煮熟成飯。

烹調竅門

怎樣才能又方便快捷地去掉栗子衣？

可將已去殼的栗子放在白鑊中炒熟，
栗子衣便容易剝出。現在市售的栗子
很多都已去衣，可省回去衣的時間。

番茄魚茸羹
Fish thick soup with tomato P.165

材料

鯇魚腩.............................. 80 克
番茄（大）........................ 1 個
薑茸.................................. 1 湯匙
上湯.............................. 500 毫升
葱（切度）....................... 1 條

魚調味

醬油.............................. 1 茶匙

湯調味

鹽.................... 適量

芡料 調勻

馬蹄粉........... 15 克
水................. 2 湯匙

湯品

做法

1. 魚腩洗淨蒸熟（約 8 分鐘），取出，濾去蒸出之魚水。

2. 放上葱度，煮滾適量油，立即瀨下魚面，再淋上調味，待凍拆肉備用。

3. 番茄用熱水燙片刻，去皮及籽，然後剁爛成茸。

4. 燒油 1 湯匙，爆香薑茸，傾下番茄茸爆片刻，注入上湯，加入適量鹽調味及芡料推勻。

5. 最後下魚肉拌勻，待再滾起即可上鍋。

烹調竅門

這魚又不是即蒸即食，為甚麼都要瀨油淋醬油？

如只是將魚蒸熟拆肉，做出來的魚羹會有腥味，用瀨油淋醬油的方法是防止有腥味。

註：

番茄含有蛋白質、維他命 B1、B2、C 及 P 等，而維他命 P 可以加強毛細血管，改善動脈硬化，因此番茄不僅可以維持健康，身體虛弱的人士也可以每天食用，增強體力。

海帶排骨湯

Pork rib soup with kelp

材料

排骨 .... 400 克
乾海帶 .... 30 克
薑 .... 2 片
清水 .... 2 公升

做法

1. 排骨洗淨飛水過冷，瀝乾備用。

2. 海帶浸透洗淨，切塊。

3. 以上材料加入清水和薑片同煮至沸滾，改以中慢火煲 2 小時即成。

烹調竅門

是否湯料同吃才有最大的效益？

這湯必需要湯料同吃才能達到最大的效益，排骨可以不吃，但海帶必定要吃。

註：

海帶有「淡乾」和「鹹乾」兩種。前者身乾質輕，後者味鹹體重，質量以淡乾為佳。

海帶含豐富的碘質，亦含多種氨基酸及維他命，可預防肥胖及降血壓。

韓國婦女在產後或生日時，必定吃海帶湯，除了因傳統習俗外，最重要的原因是它能消腫去水，將體內毒素排出。

Crucian carp soup with
chestnuts and white fungus P.166

栗子雪耳鯽魚湯

材料

新鮮栗子 .... 半斤
雪耳 .... 半兩
鯽魚 .... 1 條
陳皮 .... 1 塊
清水 .... 8 杯

調味料

鹽 .... 適量

做法

1. 栗子去衣,洗淨瀝乾。

2. 雪耳浸透,洗淨。

3. 鯽魚洗淨,抹乾用油煎至兩面微黃盛起。

4. 煲中注入清水 8 杯,加入栗子、雪耳、鯽魚及陳皮,以中慢火煲約 2 小時,加鹽調味即成。

烹調竅門

可用乾栗子代替鮮貨嗎?

原稿用半斤栗子是鮮貨,如鮮栗子不當造,可用乾栗子代替,份量 4 兩便已足夠,曬乾的栗子其營養價值是一樣的。

註:栗子味甘、性溫,能固腎健胃。

蓮藕章魚
豬手湯

Pork trotter soup with
lotus root and dried octopus P.166

材料

蓮藕 ....600 克
章魚 ....1 隻（80 克）
豬手 ....1 隻（約 900 克）
紅棗（去核）........................5 粒
清水 ....10 杯

調味料

鹽..適量

做法

1. 豬手刮淨餘毛，斬件，洗淨，飛水，過冷河備用。

2. 蓮藕原節洗擦淨；章魚、紅棗洗淨。

3. 將以上各料同置煲中，注入清水，滾後改以中慢火煲 3 小時即可加鹽調味。

烹調竅門

怎樣才可弄清豬手的毛？吃時的口感與燜有不同嗎？

購買豬手時可請肉枱代為刮淨或燒清豬毛，回家後清洗乾淨，以大滾水飛水過冷，再以小刀刮淨便可。

用豬手煲湯與燜豬手的口感大不同，煲湯的入口腍滑，燜則皮爽入味而不膩口。

龍眼牛脹湯

Beef shin soup with dried longans P.167

材料

牛脹（牛腱） .... 240 克
黃芪 .... 10 克
乾龍眼肉（桂圓） .... 20 克
清水 .... 4 杯

調味料

鹽 .... 少許

做法

1. 牛脹切厚片，飛水，過冷河，瀝乾。

2. 將黃芪及龍眼肉置煲中，加入牛脹和清水。

3. 以中慢火煮 1 1/2 小時即成，加少許鹽調味。

烹調竅門

如想吃到嫩滑的牛脹，可選購金錢脹。如將牛脹原條煲，時間較長才睑滑而且較燥火，所以建議切片後才煲。

註：這湯能恢復疲勞及腦力，對產後健忘症有幫助。

花膠椰肉燉烏雞

Double-steamed silkie chicken soup with fish maw and coconut P.167

材料

花膠 .... 80 克
鮮椰肉 .... 300 克
烏雞 .... 1 隻
紅棗（去核）............. 5 粒
滾水 .... 適量

調味料

鹽 .... 適量

做法

1. 花膠用凍水浸過夜，將水煮滾熄火 5 分鐘，放入花膠，待水全涼後，撈起花膠，將水中雜質隔去（但有透明如啫喱狀物體要保留，不可棄去，因這是第一次浸出的膠質，靚而厚的花膠必定會有），再以凍水浸 3 天（每天要換水一次，要放在雪櫃）。

2. 椰肉切塊；烏雞去肺洗淨，飛水過冷備用。

3. 將以上處理好的材料同置大燉盅內，加入隔出的膠質及浸透之花膠，注入適量的沸滾水。

4. 以大火燉約 4-5 小時，可加少許鹽調味。

烹調竅門

怎樣處理或儲存第一次浸出的花膠膠質？

第一次浸出的膠質，可略用清水沖淨，用碗盛起，可於當天煲湯、燜餸、煮菜時加入同煮，或可作燉奶及煮甜品之用。如當天未能有空烹調，可先放雪櫃貯存改天再用。

註：這湯能養顏潤膚，滋補肝腎。

清燉雞汁
Double-steamed chicken essence P.168

材料

嘉美雞或康保雞 .... 1 隻
中型湯碗 .... 1 個
小碗 .... 1 個
清水 .... 4 湯匙

做法

1. 雞去皮洗淨，起肉剁碎。

2. 將小碗反轉放入湯碗中，把剁碎之雞肉覆蓋於小碗面，然後淋上 4 湯匙清水。

3. 蒸鍋放水煮滾，將蓋有雞肉之湯碗放入鍋中，滾後以中慢火燉 4 小時。

4. 取出湯碗，移去雞肉及小碗，湯碗內的雞汁便可飲用（不可加任何調味）。

烹調竅門

如不用雞，可改用瘦肉 300 克，做法相同。燉汁後之雞肉或瘦肉，還可取出加入其他蔬果同煮湯，給其他家人食用。

註：
這肉汁主治脾虛體弱、頭暈心悸、月經不調、產後補養，尤合老人及病後調養者。如有感冒，勿飲燉雞汁改為飲燉肉汁。

雙棗黑豆燉鯉魚

Double-steamed common carp soup with red dates, black dates and black beans P.168

材料

鯉魚（不需要去鱗）	1 條
紅棗（去核）	5 粒
黑棗	5 粒
薑片	2 片
黑豆	40 克
滾水	適量

調味料

紹酒	2 湯匙
鹽	適量

做法

1. 鯉魚洗淨抹乾，用適量油煎至兩面金黃，用熱水沖淨油脂。

2. 洗淨紅、黑雙棗；黑豆以白鑊炒至破皮，然後灒入清水，滾 3 分鐘盛起過冷，瀝乾。

3. 將以上各料和薑片同置燉盅內，注入適量滾水。

4. 將燉盅轉置蒸鍋中，隔水燉約 4 小時即可調味食用。

烹調竅門

鯉魚有利水消腫、益氣旺血、溫中固腎、壯陽補虛的作用，民間有用鯉魚鰾治疝氣；鯉魚鱗片能止血，而且鱗片含魚膠成份亦多，煎後有甘香去腥作用，故鯉魚不用去鱗。燉湯適宜注入煮滾的水，但涼的開水亦可，但不宜用生水。

註：這湯能補虛利水、養血通乳。

鮑魚北芪燉竹絲雞

材料

鮮鮑魚 .... 4 隻
烏雞（竹絲雞）......... 1 隻
北芪 .... 30 克
淮山 .... 30 克
杞子 .... 10 克
陳皮（浸透）............ 5 克
桂圓肉 .... 5 粒
生薑 .... 2 片
滾水 .... 適量

調味料

鹽 .... 適量

做法

1. 鮑魚擦透，洗淨飛水過冷，去殼及腸備用。

2. 烏雞去肺洗淨，飛水過冷瀝乾。

3. 將其餘材料洗淨，與鮑魚、烏雞同置燉盅內，
   注入適量滾水，以中火隔水燉 4 小時，用適量
   鹽調味即成。

Double-steamed silkie chicken soup
with abalones and Bei Qi P.169

烹調竅門

怎樣清洗鮮活的鮑魚？

將刀沿鮑魚殼邊緣插入，切斷中間與周圍的堅硬組織，將鮑魚肉剝離，將雜質及沙袋去除，用刷子將鮑魚肉的表面黑膜擦洗乾淨，以粗鹽將附着的黏液清洗乾淨，瀝乾水便可使用。

註：這湯能滋陰養血、健脾理氣。

通草豬腳湯

Pork trotter soup with Tong Cao P.169

湯品

材料

豬腳	1 隻
通草	10 克
青葱	10 條
清水	8 杯

調味料

鹽	適量

做法

1. 豬腳斬件，洗淨飛水過冷，瀝乾。

2. 通草略洗，青葱連鬚根一起洗淨。

3. 先將豬腳、通草放入鍋中，注入清水，以大火煮至沸滾後，改用中慢火煲 2 小時，再加入帶根的青葱。

4. 再以慢火煮 10 分鐘，即可加鹽調味，飲湯食豬腳。

烹調竅門

為甚麼青葱需要連根一起煲湯？

葱根可治傷寒頭痛，還可入饌佐膳。

註：

這湯主治產後氣血不足，乳汁稀少、色淡或無乳。

葱的葱白部份含豐富的維生素 C，可以溫暖身體，具有發汗退燒作用，對改善感冒初期的症狀十分有效，葉子（綠色）部份含有可以維持黏膜健康的 B-胡蘿蔔素，能抵抗呼吸系統的感染。

葱的辛香味來自硫化丙烯，有助體內排除毒素，增進肝臟的解毒功能，也具有抗氧化及增強免疫的功效。

它具有發汗的功能，利於排汗作用，同時促進體內毒素由汗腺排出。

紅棗木瓜花生湯

Papaya soup with
peanuts and red dates P.170

材料

木瓜 .... 1 個
紅棗（去核） .... 12 粒
無花果（切片） .... 6 個
有衣花生 .... 80 克
腰果 .... 80 克
陳皮 .... 1 塊
清水 .... 10 杯

做法

1. 木瓜去皮，切塊。
2. 將以上所有材料冷水入鍋，煲 2 小時即成。

烹調竅門

為甚麼材料要冷水入鍋？為甚麼指定要有衣花生？

冷水落料，食物在凍水內慢慢加熱，對人體有益的好東西便會慢慢地融入湯水裏去，很多時高溫會使營養流失。

宜用有衣、細顆、稱為珠豆的花生，它與木瓜同煲湯飲用有催乳的功效。

註：木瓜又稱萬壽果。此湯水含有豐富蛋白質、礦物質及維他命 A、B 及 C，能養顏潤膚、鎮咳祛痰、明目清熱、清腸理胃及通便。

材料

佛手瓜（合掌瓜）......2 個
南瓜 ....300 克
番茄 ....2 個
冬菇 ....60 克
清水 ....8 杯

調味料

鹽............................適量

做法

1. 冬菇浸軟，洗淨。

2. 佛手瓜及南瓜去皮、核，切塊。

3. 番茄連皮切塊。

4. 將以上材料同置煲中，注入清水，待沸滾後，
   改以中慢火煲 2 小時，即可加鹽調味。

烹調竅門

煲蔬菜湯材料是否一定要滾水下鍋？

一般人煲老火湯愛在水滾後下材料，我卻喜歡凍
水入料，因為有些材料尤其是骨類，如水滾後才
下鍋，會把食物的表面纖維封死，使營養未能充
分釋出，便發揮不到全面的補益作用。

有人說煲菜湯若凍水落鍋，湯會苦，我未從遇過
這問題，可能是菜的本質已苦吧！

註：這湯清甜滋潤，維他命豐富。

佛手南瓜番茄冬菇湯

Chayote and pumpkin soup with
tomato and shiitake mushrooms P.170

蓮子菠菜銀耳湯

Spinach soup with
lotus seeds and white fungus P.171

材料

菠菜 ....300 克
新鮮蓮子 ....100 克
銀耳（雪耳）.....75 克
薑 ....2 片
清水 ....2 杯

調味料

鹽 ....適量

做法

1. 銀耳洗淨浸透，蓮子洗淨飛水瀝乾。

2. 菠菜洗淨（根頭部份留下），切段。

3. 鍋中注入清水，加入薑片、蓮子和銀耳煮 10 分鐘。

4. 加入菠菜，待湯汁再沸滾起，即可加鹽調味食用。

湯品
素

烹調竅門

銀耳有甚麼功效？為甚麼銀耳經煲湯後，有些會腍滑，有些則較爽口呢？

雪耳又稱銀耳，功能清肝潤肺、滋陰養顏，對於虛火上升、煩燥失眠、食慾不振、虛不受補者為補益品，對身體很有好處。銀耳在中國很多地區皆有出產，但以福建漳州的品質最好。

如想口感腍滑，可於烹調前以溫水浸泡 2 小時，待耳身全脹發透，再以冷水洗淨。如要爽口，用冷水浸發便可。

註：此湯能補中養神，除百疾。湯料含有多種營養，如菠菜含豐富的鐵質，尤其是它的根頭部；銀耳含豐富的氨基酸和多糖的膠原蛋白；而蓮子則是養生藥材，常用於食療保健，含有蛋白質、脂肪、碳水化合物、鈣、磷、鐵等成分。

鮮百合蜜豆炒肉片

Stir-fried sliced pork with
lily bulb and sugar snap peas P.172

材料

瘦肉 .... 100 克
鮮百合 .... 80 克
蜜豆 .... 80 克
濕發雲耳 .... 50 克
薑汁酒 .... 1 茶匙
蒜頭（切片）.............. 1 粒

醃料

醬油 .... 1 茶匙
生粉 .... 1/4 茶匙
熟油 .... 1 茶匙
水 .... 1 茶匙

調味料

鹽 .... 1/4 茶匙
糖 .... 1/4 茶匙
醬油 .... 1 茶匙
生粉 .... 1 茶匙
麻油 .... 少許
胡椒粉 .... 少許
水 .... 5 湯匙

小菜

做法

1. 瘦肉洗淨切片，加醃料拌勻；百合洗淨，一片片的撕開。

2. 蜜豆、雲耳飛水過冷瀝乾。

3. 燒油 1 湯匙，放下肉片，炒至熟盛起；洗淨鑊。

4. 燒油 1 茶匙，爆香蒜片，將蜜豆、雲耳回鑊，加入百合炒透，傾下肉片，灒薑汁酒，再炒至百合半透明狀，即可加入調味拌勻上碟。

烹調竅門

怎樣辟去蜜豆的豆腥味？百合需要飛水嗎？

凡有草腥味的蔬菜類，只要用蒜或薑起鑊爆透，再加適量水冚蓋片刻，然後將物料撈出，倒去鑊中的水分，便可去除腥味。

百合不能飛水，因會變色的。

冬菇紅棗蒸斑片

Steamed grouper slices with shiitake mushrooms and red dates

材料

石斑肉.............................240 克
冬菇（切片）.........................4 朵
紅棗（去核，切片）...................4 粒
薑（切絲）...........................2 片
葱（切絲）或芫茜（切段）.........1 棵

醃料

薑汁酒.............................1 湯匙
鹽.................................半茶匙
胡椒粉...............................少許
生粉.................................1 茶匙
熟油.............................1 湯匙

做法

1. 石斑肉洗淨瀝乾，切片。

2. 斑片拌入冬菇、紅棗和醃料，醃約 15 分鐘。

3. 將斑片和薑絲置碟中，放蒸籠內，以大火蒸 6 至 8 分鐘。

4. 將蒸出之魚水傾去，灑下葱絲或芫茜段，燒滾 2 湯匙油，然後淋在斑片上即成。

烹調竅門

怎樣挑選石斑肉？以及怎樣能令斑片又嫩又滑？

游水魚當然新鮮，但死去不久的仍可食用，魚販會將大魚斬件出售。要判別其新鮮程度，此時要留意切口部位，若是凸出而呈膨脹現象的，是新鮮活魚生劏；若切口部位是齊口的，魚肉不凸出，那是死魚開刀，應否購買要考慮清楚。

石斑魚是一種低脂肪、高蛋白質的上等食用魚，營養豐富，肉質細嫩潔白，類似雞肉，故有海雞肉之稱。石斑魚入饌，大多以清蒸或油浸，以取其鮮美滋味不流失。

如起魚肉炒或起片涮火鍋，可先以適量蛋白生粉拌勻，煮後的魚肉會更嫩滑。

麻油老薑雞煮酒

Rice wine chicken with
sesame oil and ginger P.173

材料

雞項 .... 半隻
老薑 .... 80 克
黑芝麻油 .... 3 湯匙
糯米酒或米酒 .... 半杯
清水 .... 半杯

醃料

鹽 .... 1 茶匙
熟油 .... 1 茶匙
生粉 .... 半茶匙

做法

1. 雞洗淨斬件，加入醃料拌勻；老薑連皮切厚片。

2. 熱鑊下麻油 3 湯匙，加入老薑炒至淺褐色。

3. 傾下雞件炒透，加入糯米酒及清水，煮約 15 分鐘即可食用（不用加調味）。

烹調竅門

為甚麼要用雞項？其食味如何？如不想買到打針雞而又想食雞項有甚麼選擇？

雞項煮熟後皮色金黃潤澤，肉質滑嫩甘香，故港人愛食雞項；雞項即未經交配的雌雞。

在 60 年代，雞農引進外國技術，於雞頸注入雌激素（俗稱肥丸，即「針雞」），使雞隻生長快及積聚脂肪，故雞隻多肉及更加嫩滑。於 80 年代有研究顯示人造雌激素可能會致癌，故政府立例禁使用。

為着健康安全又顧及食味口感，我建議選購嘉美雞的第三代皇健雞，因其由交配、繁殖、飼養，以至銷售，全部百分百在香港進行，達到國際的衛生安全標準。

註：

此食譜可改善產婦的血液循環，且能把懷孕期積存的廢物完全從體內排出。

Fried cod fish in
lemon honey sauce P.173

香煎蜜汁銀鱈魚

小菜

材料

銀鱈魚........................2 片
乾葱頭（切片）........2 粒
生粉 ....適量
酒............................少許

醃料

油 ....1 茶匙
鹽 ....半茶匙

调味料

醬油 ....2 湯匙
檸檬汁 ....1 茶匙
蜂蜜 ....1 湯匙
水 ....2 湯匙

做法

1. 銀鱈魚洗淨抹乾，以醃料塗勻醃 15 分鐘；取出，撲上適量生粉。

2. 燒熱鑊下油 2 湯匙，煎至銀鱈魚兩面金黃取出。

3. 燒鑊加油少許，爆香乾葱，灒酒，倒下調味煮滾，然後淋在魚面即成。

烹調竅門

銀鱈魚的肉質非常嫩，煎時容易爛，要怎樣才能煎得完整？

用易潔鑊以適量油慢火煎，不要心急將魚翻轉，煎至金黃並非難事。

肉桂蜂蜜醬油雞

Braised chicken in
cinnamon honey soy sauce P.174

材料

嘉美雞.................半隻（300 克）
肉桂 ....10 克
生薑（切片）.....................40 克
米酒 ....1/4 杯
紅棗（浸透，去核）...........8 粒

醃料

薑汁酒...........1 湯匙
生粉 ....1 茶匙

調味料

蜂蜜 ....2 湯匙
豉油雞醬油.......................1/4 杯
鹽....................................半茶匙
清水 ....1/4 杯
* 豉油雞醬油做法看後頁（P.100）

做法

1. 雞洗淨斬件，用醃料拌勻。

2. 燒熱鍋下油 2 湯匙，放下薑片和肉桂爆香，加入雞件炒至
   微黃，潷酒，加入調味及紅棗，以慢火燜至熟透（約 10 分
   鐘）即成。

烹調竅門

為甚麼要指定用嘉美雞，牠的肉質有何特別？

嘉美雞（雌性），與嘉美少爺雞（雄性）皆是皮下脂肪少，所以白皮（因為皮層下若有脂肪才會使皮顯現黃色）。嘉美少爺雞因為分泌雌激素較少，所以脂肪含量比雞項更低，牠最大的特色是肌肉含豐富骨膠原，蛋白質含量比雞項高，吃起來皮爽肉滑而不會有肥膩感覺。如用於產後補身或病後調養，我建議選用嘉美雞。

豉油雞做法 ────────

材料

光雞 1 隻、乾葱 2 粒、生薑 2 片、八角 2 粒、冰糖 600 克、生抽 500 克、老抽 500 克、玫瑰露酒 1 湯匙

做法：

1. 雞去肺洗淨，飛水過冷瀝乾。

2. 乾葱、生薑、八角洗淨，與冰糖、豉油同置鍋中，以慢火煮溶冰糖，灒下玫瑰露。

3. 剪去雞腳，將整隻雞放入豉油中，以中慢火浸 20 分鐘（浸約 10 分鐘，將雞翻轉）。把浸熟之雞盛起待凍，斬件供食。

4. 將浸過雞之豉油隔渣，待全凍後入回豉油樽中，作調味醬油用。浸過雞的豉油，含有濃厚的鮮味，故再做肉桂蜂蜜醬油雞時加上此醬油，便可不需要加其他調味料（例如：雞粉等）。

雞
蛋
豬
腳
薑
醋

Pork trotters,
ginger and hard-boiled eggs
in sweet black vinegar P.175

材料

雞蛋（或鹹蛋）...........10 個
豬腳（斬件）.................2 隻
薑.................................3 斤
黑米醋...........................2 斤
片糖 ....2 斤

做法

1. 雞蛋焓熟浸凍水，去殼。

2. 薑洗淨，刮去皮，吹乾；豬腳飛水過冷河。

3. 用少許油爆薑片刻。

4. 將薑放於瓦罉內，注入黑米醋，滾後以慢火煲
   半小時，加入片糖，煮至糖溶，再以慢火煲
   30 分鐘，加入豬腳煲 30 分鐘，熄火焗至明天，
   再煲滾醋，以慢火再煮半小時，加入雞蛋滾
   10 分鐘後，熄火浸至入味。

烹調竅門

1. 為何薑醋會越煲越黑？

為何煲薑醋，在放下豬腳後，豬腳會越煲越硬，而薑的色澤則越煲越黑，豬腳或豬手的皮質如橡皮般韌，醋又沒有香味，只有甜味，何解？

以上這個問題，在我的陪月班學生中，曾經出現過兩次。所以我覺得應該和讀者共同分享。

一般用甜醋煲薑，色澤多會變深，但亦不至於黑似墨汁。但如你買的甜醋是用了糖精提煉，而並非由蔗來提煉，就會出現以上的問題。糖精（即人造糖）不可加高熱，否則味道會變苦。如果酸醋是用糖精提煉便更為嚴重，色澤會越煲越黑，皮肉也會越硬，猶如橡皮，並不適合食用。

因此我建議用優質的黑米醋加入片糖來煲薑醋，除了可口及避免出現上述的問題外，更因為片糖可以消炎，幫助子宮收縮，對產婦很有益處。

2. 用豬腳或豬手煲薑醋有何分別？

豬蹄是豬腳、豬手、豬肘子、蹄膀的統稱。豬腳、豬手通常指豬腿的中關節以下的部位，前蹄是「豬手」，後蹄是「豬腳」，實際上切成塊煮爛沒甚麼分別，煲薑醋兩者可用。

註：
1. 豬蹄中含有較多的蛋白質、脂肪和碳水化合物，可加速新陳代謝、延緩身體衰老，對產婦能起到催乳和美容的雙重作用。順產的婦女，十二朝後便可進食薑醋；剖腹生產的婦女，則要等三星期後才可食用。
2. 食素的產婦，煲薑醋時減去豬腳和雞蛋便可。

小菜

魚頭雲煮酒

Fish head soup with scrambled egg,
ginger and rice wine P.176

材料

大魚頭（開邊）........... 1 個
黑豆 .... 75 克
木耳 .... 20 克
雞蛋 .... 1 個
雙蒸酒..................... 160 克
薑汁 .... 1/4 杯
薑（切絲）................. 2 片
清水 .... 4 杯

做法

1. 白鑊炒黑豆至破皮，以清水一碗注入略滾片刻，盛起，過冷河，瀝乾。

2. 木耳浸透切粗條；雞蛋打散炒熟；魚頭洗淨抹乾，用油略煎至微黃盛起。

3. 以適量油爆薑絲、木耳，注入清水4 杯，加入薑汁和酒，放下魚頭、黑豆及炒好之雞蛋。

4. 以中慢火煮至湯汁剩回 2 杯時即成（約 15 分鐘）。

烹調竅門

為甚麼要煎香魚頭？如怕湯肥膩，怎樣去油？

魚頭略煎可使湯汁乳白而帶鮮香。煎好後的魚頭如怕肥膩，可先過熱水沖去油脂。

註：

「大魚」即「鱅魚」，其頭部發達，約佔全身的五分之二，故又稱大頭魚。其頭肉與頭骨之間有一膠狀白色腦脂（普遍稱為魚雲），含豐富的鈣、磷、蛋白質及脂肪等，能治頭風、頭暈，配合黑豆和酒煮成湯，對體質虛弱、中氣不足人士非常有效。

小菜

材料

馬友魚（或任何鮮魚）........10 兩
雞蛋 ....3 個
熱水 ....1 1/4 杯
葱絲 ....少許

醃料

鹽 ....半茶匙
薑汁酒 ....1 茶匙

調味料

鹽 ....1 茶匙

做法

1. 整條鮮魚起骨，用醃料塗勻，15 分鐘後備用。

2. 雞蛋打起；調味料加入熱水中。

3. 將熱水撞入蛋液中（要邊撞邊打透）。

4. 鮮魚放在深碟中，置蒸籠以大火蒸 5 分鐘，取出將蒸出之魚水倒去，吸乾水分。

5. 然後注入蛋液，以中慢火再蒸 5 分鐘，灑上葱絲於魚面中心位置。

6. 將適量油煮滾，潷在葱絲面上即成。

烹調竅門

為甚麼要將熱水撞入蛋液中，還要邊撞邊打透？蒸蛋會更滑嗎？

蒸滑蛋要用熱水撞蛋液而非用凍水打透，因為冷的東西相對是較重的，讓物料容易下墜。當一碟蛋液放在蒸鍋內，鍋內的蒸氣和熱力到沸點，熱的蒸氣會在碟的周邊，如果用凍水打蛋液，凍的蛋液較重，會在碟的中心位置，容易令蒸蛋的四邊熟而中心未熟，如火力過大，中心就會起一深窩，四邊起了蜂巢，蛋亦變老。

原條鮮魚蒸滑蛋

Steamed fish in savoury egg custard P.176

糖醋排骨

Sweet and sour pork ribs P.176

材料

腩排 .... 480 克
薑 .... 2 片
乾葱 .... 2 粒
八角 .... 1 粒
桂皮 .... 1 片
薑汁酒 .... 1 湯匙
水 .... 2 杯

調味料

鹽 .... 3/4 茶匙
醬油 .... 1 湯匙

汁料

香醋（鎮江醋）........ 1 茶匙
茄醬 .... 2 湯匙
片糖（約半磚）........ 45 克

做法

1. 腩排洗淨抹乾，斬成 2 吋長。

2. 下油 2 湯匙，將排骨煎至兩面稍黃，盛起。

3. 以少許油爆香薑、乾葱、八角和桂皮，將排骨回鑊爆透，灒薑汁酒，加水 2 杯及調味煮滾，以慢火燜約 40 分鐘。

4. 加入汁料兜勻，再燜至汁液濃稠即成。

羅漢滑豆腐

Assorted vegetables on poached tofu P.177

材料

冬菇仔	8 朵
本菇	80 克
蜜豆	80 克
甘筍	40 克
濕發雲耳	40 克
銀芽	80 克
薑	2 片
豆腐	1 塊（約 10 兩）

調味料

蠔油	1 湯匙
醬油	1 茶匙
素雞粉	1 茶匙
麻油	少許
胡椒粉	少許
糖	半茶匙
生粉	2 茶匙
水	半茶匙

做法

1. 冬菇浸透搾乾水。

2. 蜜豆洗淨；甘筍切花。

3. 豆腐切半吋厚片，飛水片刻盛起，置碟中。

4. 燒油爆香薑片，加入各材料同炒透，灒酒，下調味煮沸，即可淋在豆腐面供食。

烹調竅門

有甚麼竅門能令豆腐在切或飛水時保持完整？

除小心落刀處理外，你可將整塊豆腐用篩盛着，然後連篩一起放入滾水中，煮約 5 分鐘後盛起，待凍片刻，然後才切，便會保持完整，亦可省卻切後飛水的工序。

美果炒雙珍

Stir-fried mushroom duo
with cashew nuts P.178

材料

炸脆腰果	80 克
鮮冬菇	80 克
草菇	80 克
西芹	120 克
甘筍花	6 片
薑	2 片
芫茜莖（切度）	2 棵
薑汁酒	1 湯匙

調味料

素蠔油	1 湯匙
醬油	1 茶匙
糖	1/4 茶匙
鹽	1/4 茶匙
麻油	少許
胡椒粉	少許
生粉	1 茶匙
清水	4 湯匙

做法

1. 鮮冬菇洗淨去蒂，飛水片刻盛起，過冷瀝乾。

2. 西芹切塊，草菇在頂部剉十字，分別飛水、過冷瀝乾。

3. 燒紅鑊下油 2 湯匙，爆香薑片，傾下所有材料，以大火炒透（除薑汁酒和腰果），灒薑汁酒，下調味推勻，最後加入腰果兜勻即上碟。

烹調竅門

要怎樣處理草菇？為甚麼要在草菇頂剉十字？

草菇不適宜存放，就算放在雪櫃裏一天，菇身便會分泌出滑潺潺的黏液不能食用。新鮮的草菇底部黏積泥垢，食用前應以小刀切去及清洗，但不可用水浸泡，因草菇本身帶有一些像發霉的臭雹味，若用水浸，氣味難除，所以在烹煮之前，必定要飛水及過冷至菇身全凍後才盛起瀝乾，以毛巾或吸水紙吸乾水分，煮時亦要以適量油爆至乾透，否則餸菜雖然煮或炒透，但稍後便會出水。

在頂部剉十字只為好看有美感，其實原粒或開邊亦可以。

清煮南瓜

Braised pumpkin in soy and oyster sauce P.179

材料

南瓜 .... 300 克
紅棗（去核）.... 3 粒
薑米 .... 1 湯匙

調味料

醬油 .... 1 茶匙
蠔油 .... 1 茶匙
鹽 .... 1/4 茶匙
清水 .... 半杯

做法

1. 南瓜去皮，切大粒。

2. 紅棗洗淨，切片。

3. 熱鑊下油 1 湯匙，爆香薑米、紅棗，倒入南瓜炒透。

4. 下調味炒勻，蓋上鑊蓋煮約 10 分鐘，至汁液收至濃稠即可上碟。

烹調竅門

哪種南瓜適合烹調這菜餚？

「南瓜」別名「番瓜」或「金瓜」，一般市場出售的多是深橘色及綠色，形態各異，有扁圓形或球形。

日本南瓜外形及顏色變化多，嫩身的南瓜，皮綠瓜肉淡黃；老身的皮和肉則呈深橙色，味香甜而質感較粉糯。中國和印度出產的蜜本南瓜體積較小，身長如葫蘆，表皮光滑，肉質香甜較腍軟。

選購甚麼品種，隨食用者之愛好吧！

小菜素

香芹炒藕片

Stir-fried lotus root with
Chinese celery P.179

116

材料

蓮藕	1 節（300 克）
唐芹	1 棵
豆干	1 件
南乳	30 克
薑（切絲）	30 克
清水	1 杯

調味料

醬油	2 茶匙
蠔油	1 湯匙
糖	1 茶匙

芡料

生粉	1 茶匙
清水	2 湯匙

做法

1. 蓮藕去皮切片，唐芹洗淨切度，豆干切片。

2. 燒油 1 湯匙，爆香一半薑絲，下豆干炒片刻，加入唐芹兜勻盛起。

3. 燒油 1 湯匙，爆香餘下之薑絲和南乳，傾下蓮藕片炒勻，注入清水及調味，燜至汁液剩回一半量。

4. 將唐芹及豆干回鑊拌勻，加入芡料推勻即可上碟。

烹調竅門

有人愛蓮藕爽脆，有人則喜歡口感粉糯，究竟要怎樣挑選？

蓮藕又分「九孔」和「七孔」藕。

九孔藕又名白花藕，身瘦長，人們形容它叫西施臂，質爽脆甜，宜炒和搾汁。

功效：清熱涼血，對膀肛炎、尿道炎及腎炎有幫助。

七孔藕又稱紅花藕，形粗短，如生吃食味澀，宜煲湯，質粉糯香。

功效：補血，女士經後或產後與紅棗及紅豆煲湯最好。

小菜 素

117

Braised shiitake
mushrooms with kelp P.177

小菜
素

材料

海帶結......................80 克
薑（切片）..................40 克
冬菇 ....160 克
紅辣椒（切絲）..............1 隻
炒香白芝麻..................2 湯匙
冰糖 ....1 粒（約 5 克）
清水 ....2 杯

調味料

醬油 ....2 湯匙
素蠔油.........4 湯匙
麻油 ....1 湯匙

做法

1. 海帶結洗淨浸水至發大（約 2 小時，要經常換水）。

2. 冬菇洗淨浸透搾乾水。

3. 燒油 2 湯匙，爆香薑片及冬菇，加入清水及 1 粒冰糖，滾後以慢火煮 30 分鐘。

4. 加入調味及海帶再燜 15 分鐘，至剩餘汁液 1/4 杯時盛起，灑上白芝麻，加入紅椒絲拌勻即成（可涼食）。

烹調竅門

海帶結那裏有售？為甚麼要打結才燜？

海帶結在國貨公司或日資超市有售。

不一定要買打結的海帶，只是美觀而矣。海帶有「淡乾」和「鹹乾」兩種。淡乾質輕、鹹乾體重。質以厚實身乾，色呈濃黑綠色或濃褐色的淡乾為佳。

冬菇扒豆苗

Stir-fried pea sprouts with
shiitake mushrooms and oyster sauce P.180

材料

冬菇 .... 40 克
豆苗 .... 300 克
薑絲 .... 1 湯匙
薑 .... 2 片
冰糖 .... 半茶匙

調味料

素蠔油 .... 1 湯匙
上湯 .... 半杯
醬油 .... 1 茶匙

芡料

生粉 .... 1 1/2 茶匙
水 .... 4 湯匙

做法

1. 冬菇洗淨浸透，搾乾水切片。

2. 豆苗洗淨瀝乾。

3. 燒油 2 湯匙，爆香薑絲，傾下豆苗炒至
   軟身，冚蓋 2 分鐘，即可盛起瀝去汁水
   置碟中。

4. 燒油 1 湯匙，爆香薑片，傾下冬菇炒透，
   加入調味和冰糖煮滾，慢火煮 2 分鐘，
   加入芡料推勻，然後淋在豆苗上。

烹調竅門

冬菇為甚麼要搾乾水分才切片？

凡是以冬菇作的菜式，都要將其浸透後搾
乾水分才切片，除方便處理所需要之形狀
外，清潔衛生最為重要。

註：

豆苗含有豐富的鈣質、維他命和胡蘿蔔
素，有利尿、止瀉及消腫作用，更能幫助
消化，常食可令皮膚光滑柔軟，面色紅潤。

菠菜木耳炒香干

Stir-fried dried tofu with
spinach and wood ear fungus P.178

材料

菠菜 ....320 克
木耳（浸透）...................10 克
五香豆干 ....1 塊
薑（剁碎）........................2 片
清水 ....1/4 杯

調味料

蠔油 ....1 茶匙
鹽............1/4 茶匙
生粉 ....1 茶匙
清水 ....2 湯匙

做法

1. 菠菜洗淨，木耳浸透切條，五香豆干切片。

2. 熱鑊下油 2 湯匙，爆香薑碎、木耳及豆干，加入菠菜炒片刻，
   注入清水 1/4 杯（約 5 湯匙），合蓋煮 2 分鐘盛起，倒去汁水。

3. 將材料回鑊，加調味兜勻上碟。

烹調竅門

一般人認為菠菜多食有益，是否對人體所有機能都有補益功効？

菠菜能解酒毒、除胸膈翳悶、清血熱治皮膚痕癢，唯是從中國古
傳食療一書中得悉，吃菠菜太多或長期食用會損害腎臟，如果腎
機能有問題之人，例如性神經衰弱、夢遺早洩、滑精腰痛、婦女
經期過多、小腹寒冷、夜尿頻密等症狀，是不宜多吃菠菜的，尤
其是脾腎衰弱，吃了過多菠菜或飲菠菜湯，往往會引起腹瀉，如
有慢性腸膜炎疾患的人，更不宜食用。

做好的炒米，可配任何食療材料使用，例如：陳皮、紅棗、通草或北芪等，煮好的茶水是代替水飲用，記着炒米茶必須每日新煮，如飲剩必需棄去。

材料

白米 .... 400 克

做法

用白鑊炒白米至金黃色，待凍，以瓶盛起備用。

炒米

Toasted rice P.181

炒米茶

Toasted rice tea P.181

材料

清水 .... 12 杯
炒米 .... 3 湯匙
陳皮 .... 1 角

做法：

1. 煲中注入清水，加入陳皮及 3 湯匙炒米（不用清洗）。

2. 待煲至炒米開花，便可以保溫瓶盛起，可供產婦全日飲用。

材料

紅棗（去核）...... 20 粒
生薑 .... 2 片
炒米 .... 2 湯匙
清水 .... 12 杯

做法

1. 紅棗洗淨置煲中，加入炒米及薑片。

2. 注入清水，待沸滾改以慢火煲 1 小時即成。

註：這茶有祛風暖胃的功效

紅棗炒米茶

Toasted rice tea with red dates P.182

材料

通草 ....10 克
北芪 ....15 克
炒米 ....2 湯匙
清水 ....12 杯

做法

將以上材料洗淨，放入煲內，注入清
水煮至沸滾後，改以慢火煲 1 小時至
炒米開花即成。

通草北芪茶
Tong Cao tea with Bei Qi P.182

材料

桑寄生.....................120 克
雞蛋.......10 個（或適量）
冰糖.........................適量
清水.....................12 杯

做法

1. 雞蛋煮熟，浸凍水去殼
   備用。

2. 桑寄生洗淨放煲內，加
   入清水煲滾後，以中火
   煲 20 分鐘，放入雞蛋再
   煲 40 分鐘，撈起桑寄生
   棄去。

3. 下冰糖，煮至糖溶即成。

註：桑寄生蛋茶有養血祛
風、強肝健體、補益筋骨
之功效。

桑寄生蛋茶

Sang Ji Sheng tea with
hard-boiled eggs P.183

紅棗合桃露

Creamy ground walnut sweet soup P.183

材料

合桃肉.........................2 杯
紅棗（去核）.............8 粒
白米 ....半杯
冰糖 ....320 克
清水 ....8 杯

做法

1. 合桃肉洗淨，用熱水浸透。

2. 紅棗用熱水略煮，去核備用。

3. 白米洗淨浸 4 小時。

4. 把以上之材料放攪拌機中，加入清水 4 杯攪成糊狀，過濾後放入鍋中（不可用鐵鍋）。

5. 再用 4 杯清水將冰糖煮溶，加入合桃糊內，一邊煮一邊攪拌，滾後便成。

烹調竅門

如合桃露太稀，有甚麼補救方法？為甚麼不可用鐵鍋？

如攪拌機太細，可將以上材料分兩次攪拌。如煮成糊後覺得太稀，可用粘米粉開水，用打芡方法加入糊中煮至稠。

如用鐵鍋煮合桃露，會有苦澀味。

甜品篆

鮮奶燉木瓜

Double-steamed papaya sweet soup with milk P.184

材料

木瓜 .... 1 個
冰糖 ......2 湯匙（或隨意）
鮮奶 .... 1 杯
清水 .... 1 杯

做法

1. 木瓜洗淨，去皮核，切塊，置燉盅內。

2. 冰糖放入清水中煮溶，注入盛載木瓜之燉盅內。

3. 以大火燉 3 小時，加入鮮奶再燉 30 分鐘即成。

烹調竅門

市上有綠色及橙色的木瓜，那種最宜用作燉木瓜？

樹上熟透的木瓜，可以作生果食用，至於未成熟的木瓜可用以燜魚燴肉及煲湯用，半熟的木瓜用作燉奶最適宜。

註：

燉木瓜能治咳嗽、潤肺，更有養顏之功效。

鮮奶燉木瓜全家老幼皆可享用，故這食譜的材料份量會比較多，約 2 至 4 人用。

材料

糯米 .... 200 克
蓮藕 .... 1.28 千克
白糖 .... 200 克
桂花糖 .... 適量

做法

1. 糯米浸過夜，洗淨瀝乾。

2. 蓮藕洗淨，在前端近藕節處切一小蓋。

3. 將浸透之糯米灌入藕孔內，然後蓋上，以牙簽固定藕蓋。

4. 將蓮藕放鍋中，注入過面清水，滾後以中慢火煲 2 小時。

5. 加入白糖與桂花糖同煮約 1 小時，待煮成稀糖膠狀，即可盛起待凍切片。

6. 將原汁澆在藕片上即成。

烹調竅門

藕孔內的淤泥怎樣清洗？要怎樣挑選蓮藕才沒有淤泥？

藕孔內的淤泥可以用水喉對着藕孔沖淨，如用作煲湯或燜煮，可切開來洗。但做釀蓮藕便要買有藕節的為佳，因有藕節不會容易入淤泥，並且糯米亦不會從另一面流出。

桂花糖釀蓮藕

Lotus root stuffed with glutinous
rice in candied osmanthus syrup P.184

鲜百合南瓜露

Pumpkin sweet soup with coconut milk,
sago and lily bulbs P.185

材料

南瓜	320 克
西米	60 克
糖	160 克
椰汁	1 杯
鮮百合	80 克
清水	6 杯

芡料

粟粉	2 湯匙
水	3 湯匙

做法

1. 南瓜開邊去籽，蒸腍，取出瓜肉壓成茸備用。

2. 百合洗淨，一片片剝瓣，備用。

3. 西米用水煲滾，邊煮邊攪，煮至中心點呈一點小白粒時撈出，用凍水沖去膠質備用。

4. 煲滾水約 6 杯，倒入南瓜茸、糖及西米，煮滾，加入芡料攪透勾芡，加入椰汁及百合滾片刻即成。

烹調竅門

西米易煮難精，有甚麼竅門呢？

煮西米待不用冚蓋，冚蓋會滾瀉。

煮前，將浸透的西米瀝去水分，然後放入大滾水中煮片刻，中心便呈現一點小白粒，這表示已經熟了，再熄火冚蓋焗片刻便成。如煮至全透明，就會變成漿糊狀，不好吃了。

腰果露

Creamy ground cashew nut sweet soup P.185

材料

腰果 .... 160 克
冰糖 .... 220 克
水 .... 5.3 量杯
粟粉 .... 30 克
鮮奶 .... 120 克

做法

1. 腰果洗淨，吸乾水分，置焗爐焗至金黃盛起。

2. 腰果放在攪拌機內，加入水 1 杯，磨幼，用幼篩過濾。

3. 把磨好的腰果與 4.3 杯水同置煲中煮滾，加入冰糖煮至糖溶。

4. 粟粉用少許水開勻，加入糖水中打勻，最後加入鮮奶，即成香滑之腰果露。

烹調竅門

為甚麼要最後才下鮮奶？

做甜品如要加入鮮奶，最後才下的好處是口感夠滑，而且奶味香濃。

材料

紅棗 ....6 粒
糯米粉 ....80 克
粘米粉 ....1 1/2 湯匙
清水 ....120 克
酒釀 ....50 克
冰糖水 ....1 杯
桂花糖 ....1 茶匙

做法

1. 糯米粉、粘米粉拌勻，加入清水搓成軟滑粉糰。

2. 冰糖水和紅棗一起煮滾。

3. 將粉糰搓成小丸子，放入滾水中煮熟撈出，放在煮滾之紅棗冰糖水中。

4. 再把酒釀、桂花糖加進糖水內即成。

烹調竅門

加入粘米粉搓丸子有甚麼作用？

凡搓糯米丸子（湯丸或糯米糍），加入粘米粉的好處是不黏口，入口有軟糯而帶 Q 的質感。

江南桂花酒釀丸子

Glutinous rice balls in candied
osmanthus syrup with distillers grains P.186

桂圓杏仁茶

Almond milk with dried longans P.186

142

材料

南杏 .... 120 克
北杏 .... 40 克
白米 .... 100 克
冰糖 .... 240 克或適量
桂圓 .... 20 克
清水 .... 8 杯

做法

1. 南北杏、白米洗淨,用水浸過夜。

2. 將 8 杯清水分次與南北杏、白米同置攪拌機中,磨成杏仁米漿,隔渣後備用。

3. 將冰糖舂碎;桂圓洗淨切幼粒,再把兩者與杏仁米漿同置煲中,以慢火煮滾至糖完全溶解,即可享用。

烹調竅門

白米有甚麼作用?可用粟粉代替嗎?

白米即「粳米」亦稱「粘米」,種於水田,少膠質易消化,人體所需要之糖分可從米吸收。用白米磨成粉(即粘米粉)可作糕餅,或用水開成米漿作打芡用,對腸胃合宜。如用作打芡,粟粉代替亦可。

薑汁蛋白燉鮮奶

Double-steamed milk
custard with ginger juice P.187

材料

清水	半杯
砂糖	2/3 杯
薑汁	2 湯匙
蛋白	160 克
鮮奶	250 克

做法

1. 先用清水煮溶糖，量出 125 毫升糖水加入薑汁攪勻。

2. 蛋白打勻，加入鮮奶再打透。

3. 將熱糖水撞入蛋白奶內，邊落邊打透，用布隔去泡後，將蛋白奶分放在碗內，轉置蒸籠，冚蓋，以中火蒸約 30 分鐘即成。

這是我的蒸物好拍檔——竹蒸籠，它沒有倒汗水的弊病，非常好用。

烹調竅門

用熱糖水撞入蛋白奶內，有甚麼好處？可以加入燕窩嗎？

用熱糖水撞入蛋白奶中並不停打勻，就算蒸蛋過時質感也不會老。

如要加入燕窩，可在煮糖水時加入同煮，以上份量可加濕發燕窩 40 克。

補血 Pork liver soup P.187
豬膶水

材料

豬膶（豬肝）...................... 150 克
薑（切絲）........................... 1 片
薑汁 .... 1 湯匙
米酒 .... 1 湯匙
清水 .... 1 杯

做法

1. 豬膶洗淨，剁爛成醬，盛起。

2. 加入薑汁、米酒及薑絲拌勻。

3. 清水煮滾，傾下拌勻之豬膶料，滾至豬膶熟透，
   隔渣飲用。

註：可加入少許鹽調味。

Pork rib soup with
white radish and lotus root P.188

蘿蔔蓮藕排骨湯

材料

連皮白蘿蔔（搾汁） .... 20 克
連皮蓮藕（頭尾兩邊都要有藕節）.... 1 節
豬排骨 .... 400 克
清水 .... 8 杯

調味料

鹽 ....少許

做法

1. 排骨洗淨飛水，蓮藕連皮洗淨同置煲中。

2. 加入蘿蔔汁及清水。

3. 煮滾後以中慢火煲 3 小時，加少許鹽即可食用。

烹調竅門

為甚麼一定要有藕節？

藕節可止血散瘀。藕節甘能補中、鹹能軟堅去瘀、澀能斂散固精。藕節同蓮藕一樣可治便秘，促進有害物質排出，降低血脂和血糖，具有預防糖尿病和高血壓的作用。如此好的東西，要保留不要棄去。

註：這湯能預防脹氣，吃的份量與時間可隨意。

食療

鋁箔鹽巴烤鮮橙
Roasted orange with salt (P.188)

材料

鮮橙 .... 1 個
鹽 .... 少許
鋁箔紙（錫紙）........ 1 張

做法

1. 將橙沿果蒂部位切開，露出果肉，然後灑少許鹽。

2. 把切下的部份再蓋回橙肉上，以鋁箔紙把整個橙包好。

3. 放焗爐內烤焗 20 分鐘（趁熱去皮吃橙肉）。

烹調竅門

如家中沒有焗爐怎辦？

如沒有焗爐，用多士爐亦可。倘若家中有舊的鑊或刮花了的易潔煎鍋（最好有 3 吋高或以上），請不要棄去，因鋪上錫紙便可當焗爐使用。

註：烤鮮橙有治感冒、化痰、助消化、通便等作用。

活血補虛

生化湯

Sheng Hua Soup P.189

152

活血補虛，排清惡露

根據古書記載，生化湯以活血為主，普遍用於婦女產後補血、排惡露，更可以提高抗體力量，也對子宮收縮有明顯的幫助。但為何現今的產婦不肯接受這在深厚傳統中醫藥膳精研出來的補品呢？

無論讀者接受與否，好的東西我必定介紹，至於要不要喝，就用閣下的聰明智慧來決定，或多請教德高望重的中醫師。

請珍視坐月的傳統智慧，生產是女性一生之中最大的轉變機會，應好好地把握，補好身體，這樣才可過健康而調和的生活。

中醫師的建議

生化湯是產婦在分娩後立即要喝的補品，不論是順產或剖腹生產都可以飲用。服用法如下：

順產的產婦可以在第一天開始服用，或在嬰兒出生後頭七天中服用（每天一服，連續服兩天）便可。

如動手術剖腹生產，要待三星期後才可服用。

生化湯的標準材料

當歸身	8 錢
黨參	一兩半
山藥	4 錢
炙甘草	2 錢
川芎	3 錢
白朮	3 錢
炒白芍	4 錢
薑炭	3 錢
桃仁	3 錢
生地	3 錢
熟地	5 錢
雲苓	5 錢
蒲黃	3 錢（包煎）
五靈芝	3 錢

做法

將所有材料和 5 1/2 杯水放入瓦煲內，用中火煲至剩 1 杯水即成。

注意的事項

· 如產婦有感冒病徵，就不要服生化湯，一定要看醫生。

· 盡量避免在產後第一週洗頭。

· 甜食可以酒釀為主。

** 以上寶貴的資料，由盧壽如中醫師提供。

產後出現的症狀問題

產婦患上傷風感冒，可以吃補品嗎？

產婦患上傷風感冒一定要看醫生，中、西醫都可以，不能隨意服成藥，在坐月期間，產婦身體較虛弱，最容易患上感冒，在患病期不能進食所有補品，清燉雞汁更不能喝，如必定要吃，只可以用瘦肉汁來代替雞汁。

產婦如有筋骨酸痛怎辦？有吃的禁忌嗎？

產婦因餵奶時間長，坐月期內常抱着嬰兒餵奶，維持同一坐姿太久，又常於室內抱着嬰兒走動，哄他入睡，所以產生筋骨酸痛的現象。如想解決以上問題，坐月內的餵奶正確姿勢，是應該趴着或側躺，如此骨盆與腰椎才不會因受力太重而彎曲及疲勞，產婦應盡量臥床休息，經常走動和餵奶姿勢不對，都會造成日後腰酸背痛的原因。

糯米是產後最佳補品之一，但如產婦有筋骨酸痛的情況出現，就不能吃糯米及用糯米搭配的任何食物。

產後憂鬱有食療嗎？

以中醫師角度，產婦因產後疲勞，肝臟不舒暢，才會患上憂鬱症，如果疲勞消除了，肝臟功能回復正常，就不會有此病況出現。因此，我們最好去看看中醫師，開些適合產婦個人體質，舒肝、活血及養血的藥膳湯水，憂鬱症自然消除，如果今日疲勞能今日消除，則不僅不會留下憂鬱症，有時連產前的感冒、失眠、皮膚病、氣喘、骨頭酸痛等症狀也會不藥而癒，若能於飯前例行下床動一動，很有助益。

產後減肥勿操之過急

　　一般女性為了盡快恢復產前的身材，於是產後不久就馬上做運動及走路來恢復體型，而飲食方面也限制不敢多吃，以免發胖，這實在是大錯特錯的想法。因為產後抵抗力減弱，易患感冒，疲勞難以消除，因此調養是最重要的，若是休養不足，就會種下病根，將來很容易未老先衰等，百病叢生。但亦切忌大補特補，既浪費亦損健康，滋補過量容易致肥，會使新陳代謝失調，使奶水中的脂肪含量增加，若嬰兒的消化力弱，吸收不好，會出現腹瀉，還會造成嬰兒營養不良。

　　如想皮膚和體形回復少女時代，產婦們就要好好地休息，讓子宮盡快恢復原狀，將子宮中的穢物及污血完全的排出，如子宮恢復為真空狀態，其功能也會比懷孕前更好，能調節體內荷爾蒙的代謝，同時亦有助血液及淋巴液的循環；亦因此，產婦的身體將會產生的變化是：面色紅潤、皮膚有光澤及富有彈性，更勝少女時期。

　　所以在坐月期間跟循餐單進食，不論產前是肥或瘦，經過這樣的大調理後，全面改善體質與各類宿疾，希望各位產婦能好好掌握這個能逆轉健康的良機。

Steamed pork dumplings with spinach and wood ear fungus

Ingredients

20 g wood ear fungus
(soaked in water until soft)
150 g spinach
150 g ground pork
10 g grated ginger
150 g white dumpling wrappers

Marinade

1 tbsp oyster sauce
2 tsp soy sauce
1/4 tsp salt
2 tbsp caltrop starch
2 tsp sesame oil

Method

1. Soak wood ear fungus in water until soft. Rinse and finely chop it.
2. Rinse spinach in water. Drain and blanch in boiling water briefly. Drain and rinse in cold water. Squeeze dry and finely chop it.
3. Put ground pork into a mixing bowl. Put in marinade and mix well. Add wood ear fungus, spinach and ginger. Stir again. This is the filling.
4. Lay flat a dumpling wrapper and wrap in some filling. Fold the wrapper to seal well and shape into a dumpling. Grease a steaming plate. Arrange the dumplings on top. Steam in a steamer for 10 minutes until cooked through. Serve.

Ground pork congee with red dates and pumpkin

Ingredients

300 g pumpkin
100 g ground pork
120 g white rice
10 cups water
4 red dates
(de-seeded and sliced)

Marinade

1 tsp soy sauce
1/2 tsp caltrop starch
1 tsp cooked oil

Seasoning

salt to taste

Method

1. Peel the pumpkin and dice it. Add marinade to the ground pork. Stir well.
2. Rinse the rice and put it into a pot. Add water. Bring to the boil over high heat. Turn to medium heat and cook until the rice turns mushy. Add red dates and pumpkin. Keep on cooking until creamy. Add ground pork and stir well. Cook for 10 more minutes. Season with salt. Serve.

Remarks

This congee is thicker in consistency than usual. This is actually the right consistency for the novice moms. It is a great replacement of steamed rice.

Chicken congee with dried scallops and fish maw

Ingredients

1 dried scallop
1/2 Kamei chicken
 (local organic chicken, dressed)
100 g re-hydrated fish maw
10 g shredded ginger
150 g white rice
10 cups chicken stock

Seasoning

salt to taste

Method

1. Rinse the dried scallop. Dice the fish maw.
2. Rinse the chicken and blanch in boiling water. Drain. Put chicken and water into a soup pot. Boil over high heat. Turn to low heat and cook until the soup reduces to 10 cups.
3. Remove chicken from the soup. De-bone and cut into pieces. Strain the chicken soup.
4. Rinse the rice and put into a pot. Pour in chicken soup and add the dried scallop. Bring to the boil. Turn to medium-low heat and cook until the rice turns mushy. Put in fish maw. Keep cooking until creamy.
5. Add chicken and shredded ginger. Cook briefly. Season with salt. Serve.

Soup noodles with chicken and ham

Ingredients:

1 shallot (sliced)
1/2 Kamei chicken
(local organic chicken, dressed)
40 g Jinhua ham
4 cups chicken stock
200 g Yangchun noodles

Seasoning:

salt to taste

Method:

1. Heat a wok and add a little oil. Stir-fry shallot until fragrant.
2. Rinse the chicken and fillet the chicken breast. Steam the chicken breast until done. Tear into fine shreds. Blanch the remaining chicken carcass and meat. Drain and set aside.
3. Steam Jinhua ham for about 8 minutes until done. Slice it. Then finely shred two slices and set them aside. Use the remaining sliced ham in the stock.
4. Pour 8 cups of water into a pot. Put in the blanched chicken carcass and meat from step 2 and the sliced ham from step 3. Bring to the boil. Turn to medium-low heat. Cook until the liquid reduces to 4 cups. Strain the stock and set aside.
5. In another pot, boil some water. Blanch the noodles until cooked through. Rinse in cold water. Drain.
6. Pour chicken stock from step 4 into a clay pot. Season with salt. Put in the noodles from step 5. Cook over high heat for 5 minutes for flavours to infuse. Arrange fried shallot, shredded chicken and shredded ham on top. Serve.

Congee with dried tangerine peel and cane sugar

Ingredients

120 g white rice
1 raw cane sugar slab
1 small piece dried tangerine peel
8 cups water

Method

1. Rinse the rice and put into a pot. Add dried tangerine peel and water.
2. Bring to the boil. Turn to medium heat. Cook until the rice turns mushy. Add raw cane sugar slab. Keep on cooking until creamy. Serve.

 ## Congee with dried longans, goji berries and red dates

Ingredients

15 g dried longans
1 tbsp dried goji berries
6 red dates (de-seeded)
120 g white rice
8 cups water

Method

1. Rinse the rice. Rinse dried longans, goji berries and red dates separately.
2. Put all ingredients (except goji berries) into a pot. Bring to the boil and turn to medium-low heat. Cook until the rice is mushy and creamy.
3. Add goji berries. Cook for 15 more minutes. Serve.

Remarks

From Chinese medical point of view, this congee is effective in regulating the Heart meridian, calming the nerves, benefitting the Spleen and promoting blood cell regeneration. It cures Heart-Asthenia and Spleen Asthenia, dizziness, palpitation, poor memory and insomnia.

Clay pot rice with frog, day lily flowers and cloud ear fungus

Ingredients

2 frogs
10 g day lily flowers
10 g cloud ear fungus
3 red dates
 (de-seeded, sliced)
2 slices ginger (shredded)
240 g white rice
280 g water

Marinade

1 tsp soy sauce
1 tbsp oyster sauce
1/4 tsp salt
1 tbsp ginger juice wine
2 tsp caltrop starch
1 tbsp cooked oil

Method

1. Dress the frogs. Rinse well and chop into chunks. Add marinade and mix well.
2. Soak day lily flowers and cloud ear fungus separately in water until soft. Rinse and squeeze dry. Tie each day lily flower into a knot.
3. Put the frogs, day lily flowers and cloud ear fungus into a bowl. Stir in shredded ginger and red dates. Mix well.
4. Rinse the rice and drain. Put into a clay pot. Add water and cook over low heat for 15 minutes until no more water is visible on the rice.
5. Arrange the frog mixture from step 3 over the rice. Cover the lid and cook over low heat for 15 minutes. Turn off the heat and leave the lid on. Let it sit for 5 more minutes. Serve.

Remarks

Frogs are potent tonic that replenishes central Qi (vital energy). It is suitable for those with general weakness and low spirits. Women who suffer from poor digestion or excessive stomach acid after childbirth would particularly benefit from this recipe.

Steamed rice with sliced fish and fish stock

Ingredients

1 crucian carp
1 tbsp shredded dried tangerine peel
(soaked in water till soft and shredded)
240 g white rice
260 g crucian carp stock

Marinade

1 tbsp cooked oil
1 tsp caltrop starch
salt

Seasoning

1 tsp salt

Method

1. Dress the fish and rinse well. Fillet and slice.
2. Put sliced fish into a mixing bowl. Add marinade and shredded dried tangerine peel. Mix well.
3. Fry the fish bones, head and belly in a little oil until golden. Add enough water and bring to the boil. Cook until it reduces to 260 g. Add 1 tsp of salt. Mix well. Strain and set aside.
4. Rinse the rice and drain well. Put into a deep steaming bowl. Pour in the fish stock from step 3. Transfer into a steamer. Steam over high heat for 20 minutes until the rice is cooked.
5. Arrange the marinated sliced fish over the cooked rice. Steam for 5 more minutes. Serve.

Remarks

Crucian carp is Warm in nature and does not lead to toxin accumulation. Being a highly nutritious food item, it is rich in protein, fat, calcium, iron and phosphorus. From Chinese medical point of view, crucian carp promotes diuresis and expels Dampness. It is highly effective in curing Spleen- and Stomach-Asthenia, oedema, bronchitis, asthma and diabetes.
It is a universal folkloric tradition in different regions of China to feed women with crucian carp soup after childbirth, because it stimulates breast milk production.

Steamed rice with yellow eel

Ingredients

320 g yellow eels
(also known as Asian swamp eels)
2 slices ginger (finely shredded)
2 sprigs spring onion
 (finely shredded)
240 g white rice
260 g water

Seasoning

1 tbsp ginger juice wine
1 tbsp soy sauce
1 tbsp oyster sauce
sesame oil
ground white pepper
1/4 tsp salt
1 tsp caltrop starch
1 tsp cooked oil

Method

1. Dress the eels. Blanch in hot water briefly. Rub and rinse off the slime. Cut into thick strips.
2. Put eels into a mixing bowl. Add seasoning and shredded ginger. Mix well.
3. Rinse the rice. Drain and put into a steaming bowl. Add water. Steam for about 20 minutes until the rice is cooked. Arrange marinated eel evenly on the rice. Cover the lid and steam for 8 more minutes. Drizzle with soy sauce and sprinkle with more shredded spring onion on top. Serve.

Remarks

This rice is said to be effective in replenishing Qi (vital energy), promoting blood cell regeneration, expelling Wind-Dampness, and alleviating Asthenia.

Egg fried rice with dried shrimps, ginger and sesame oil

Ingredients

20 g dried tiny shrimps
20 g ginger
2 eggs
2 bowls steamed rice
2 tbsp sesame oil

Seasoning

salt to taste

Method

1. Rinse the dried shrimps. Drain. Peel and rinse ginger. Chop ginger into fine dices. Whisk the eggs.
2. Heat a wok and add 2 tbsp of oil. Stir-fry ginger and dried shrimps until fragrant.
3. Pour whisked eggs on the rice. Then put the rice into the hot wok. Stir over high heat until the rice grains separate from each other. Season with a pinch of salt. Serve.

Remarks

From Chinese medical point of view, dried tiny shrimps are warm in nature. They are rich in calcium and magnesium and they make a great health tonic for those with general weakness and those recovering from sicknesses. They are most suitable for those middle-aged and elderly, pregnant women, novice mothers, and those suffering from muscle cramps in the calves due to calcium deficiency.

Japanese clam rice

Ingredients

1 kg water
1 small piece dried kelp (10 g)
200 g shelled clams
10 g wakame
  (Japanese seaweed, finely chopped)
600 g mixed rice (brown rice, red rice, white rice)

Seasoning

2 1/2 tbsp soy sauce
1 tsp salt
2 tbsp mirin (Japanese sweet cooking wine)
3 tbsp rice wine
1 tbsp ginger juice

Method

1. Put kelp to water in a pot. Bring to the boil. Add clams and cook for a few minutes, take out. Strain the stock. Discard the kelp.
2. Stir seasoning into the clams and mix well. Leave them for 30 minutes. Remove the clams and set aside. Save the seasoning for later use.
3. Pour the seasoning from step 2 into the kelp stock from step 1. Mix well. Strain the stock. Set aside 675 ml of stock to make rice. Soak wakame into the remaining stock until flavours are infused.
4. Drain the wakame and squeeze dry.
5. Rinse the rice. Rub it in water twice or three times. Drain. Let it sit for 30 minutes. Pour in 675 ml of kelp stock from step 3. Turn on the rice cooker and let it complete the cooking cycle.
6. Put clams and wakame over the cooked rice. Cover the lid and leave it on "keep warm" mode for about 10 minutes. Stir to mix the rice well with all ingredients. Serve.

Chestnut rice

Ingredients

240 g white rice
280 g blanching soup of chestnuts
18 shelled chestnuts
300 g water (for blanching chestnuts)

Method

1. Rinse the chestnuts and put into a pot. Add about 300 g of water to cover. Cook over medium-low heat for 20 minutes until the water reduces to 280 g. Set aside the chestnuts. Keep the blanching soup for later use.
2. Rinse the rice and drain. Put the rice into a rice cooker. Pour in the blanching soup of chestnuts from step 1. Arrange the blanched chestnuts on top. Turn on the rice cooker and let it complete the cooking cycle. Fluff the rice and serve.

Pork liver rice

Ingredients
160 g pork liver
240 g white rice
280 g water
2 slices ginger (finely shredded)

Seasoning
1 tbsp ginger juice wine
1 tsp soy sauce
1 tsp oyster sauce
1 tbsp cooked oil
1 tsp caltrop starch
1 tsp sesame oil
ground white pepper

Method
1. Rinse and slice the pork liver. Add seasoning and mix well.
2. Rinse the rice and put into a rice cooker. Add water and turn on the cooker.
3. When the rice almost dries out, arrange the pork liver over the rice. Cover the lid and let the cooker complete its cooking cycle. Serve.

Pork rib soup with kelp

Ingredients:
400 g pork ribs
30 g dried kelp
2 slices ginger
2 litres water

Method:
1. Rinse the ribs and blanch in boiling water. Drain. Rinse in cold water. Drain again.
2. Soak kelp in water until soft. Rinse well. Cut into chunks.
3. Put all ingredients into a pot. Bring to the boil. Turn to medium-low heat and cook for 2 hours. Serve.

Remarks:
Dried kelp comes in salted and plain varieties. Plain dried kelp is lighter in weight and drier in texture. Salted dried kelp is denser and salty in taste. Plain dried kelp is usually of a better quality.

Kelp is rich in iodine, amino acids and vitamins. It helps prevent obesity and reduce blood pressure. Korean women consume kelp soup on their birthdays and after childbirth. Apart from adhering to the tradition, they do so because kelp helps ease oedema and promotes diuresis. Excessive water in the tissue and the toxins therein are believed to be eliminated that way.

Fish thick soup with tomato

Ingredients

80 g grass carp belly
1 large tomato
1 tbsp grated ginger
500 ml stock
1 sprig spring onion (cut into short lengths)

Seasoning for fish

1 tsp soy sauce

Seasoning for soup

salt to taste

Thickening glaze (mixed well)

15 g water chestnut starch
2 tbsp water

Method:

1. Rinse and steam the grass carp belly for 8 minutes until done. Drain any liquid on the dish.
2. Put spring onion over the fish. Heat some oil until hot. Sizzle on the spring onion. Drizzle with soy sauce. Let cool and de-bone.
3. Blanch tomato in boiling water briefly. Peel, de-seed and finely chop.
4. Heat 1 tbsp of oil in a wok. Stir-fry ginger until fragrant. Put in the chopped tomato. Stir briefly over high heat. Pour in stock. Season with salt and stir in the thickening glaze. Keep stirring and cooking until it thickens.
5. Put in the de-boned fish from step 2. Bring to the boil. Serve.

Tips:

Tomato is rich in vitamins B1, B2, C and P. Vitamin P strengthens capillary blood vessels and improves arteriosclerosis. Tomato keeps us healthy and even those with compromised health should also eat tomatoes daily to energize the body.

Crucian carp soup with chestnuts and white fungus

Ingredients

300 g fresh chestnuts
20 g white fungus
1 crucian carp
1 piece dried tangerine peel
8 cups water

Seasoning

salt to taste

Method

1. Peel the chestnuts. Rinse and drain.
2. Soak white fungus in water until soft. Rinse and drain well.
3. Dress the crucian carp. Wipe dry. Fry in a wok with a little oil until both sides golden.
4. In a pot, pour in 8 cups of water. Add chestnuts, white fungus, crucian carp and dried tangerine peel. Bring to the boil and turn to medium-low heat. Cook for 2 hours. Season with salt. Serve.

Tips

Chestnuts are sweet in taste and warm in nature. From Chinese medical point of view, they secure Jing (essence of life) in the Kidneys while strengthening the Stomach.

Pork trotter soup with lotus root and dried octopus

Ingredients

600 g lotus root
1 dried octopus (about 80 g)
1 pork trotter (about 900 g)
5 red dates (de-seeded)
10 cups water

Seasoning

salt to taste

Method

1. Scrape off any hair on the pork trotter. Chop into pieces. Rinse and blanch in boiling water. Rinse in cold water. Drain and set aside.
2. Keep the lotus root at its internode length (i.e. cut only at the narrow intersections). Scrub each piece well. Rinse again. Rinse the red dates and dried octopus.
3. Put all ingredients into a pot. Add water. Bring to the boil. Turn to medium-low heat. Cook for 3 hours. Season with salt. Serve.

Beef shin soup with dried longans

Ingredients:
240 g beef shin
10 g Bei Qi
20 g dried longans
(shelled and de-seeded)
4 cups water

Seasoning
salt to taste

Method:
1. Slice the beef shin thickly. Blanch in boiling water. Rinse in cold water. Drain.
2. Put Bei Qi and dried longans into a pot. Add beef shin and water.
3. Bring to the boil. Turn to medium-low heat and cook for 90 minutes. Season with salt. Serve.

Remarks:
This soup helps recuperate physically and mentally exhausted. It is good for those women with poor memory after childbirth.

Double-steamed silkie chicken soup with fish maw and coconut

Method:
1. Soak fish maw in cold water overnight. Drain. Boil a pot of water and let it cool off for 5 minutes. Put in the fish maw and cover the lid. Leave fish maw in the water until the water turns cold completely. Set aside the fish maw. Check the water for any scum or dirt. (But if you find any gelatinous substance in the water, keep it for later use. Do not discard it. It is the collagen from the fish maw and only quality fish maw yields such gelatinous substance.) Soak the fish maw in cold water for 3 days and keep it in the fridge. Replace the water once every day.
2. Cut the coconut into chunks. Remove the lungs of the silkie chicken. Rinse well. Blanch in boiling water. Rinse in cold water. Drain and set aside.
3. Put the ingredients into a large double-steaming pot. Put in the gelatinous substance you get from blanching the fish maw. Add re-hydrated fish maw. Pour in boiling water to cover.
4. Double-steam for 4 to 5 hours over high heat. Season with salt. Serve.

Ingredients:
80 g dried fish maw
300 g shelled coconut
1 silkie chicken
5 red dates (de-seeded)
boiling water

Seasoning:
salt to taste

Remarks:
This soup smooths and moistens the skin, while nourishing the Liver and Kidneys.

Double-steamed chicken essence

Ingredients

1 Kamei chicken or Honbo chicken
(local organic chicken, dressed)
1 medium-sized soup bowl
1 small bowl
4 tbsp water

Method

1. Skin the chicken and rinse well. De-bone and finely chop the flesh.
2. Turn the small bowl upside down and put it into the soup bowl. Spread the chopped chicken meat over the small bowl. Pour 4 tbsp of water over the chicken.
3. Boil a pot of water. Put the soup bowl together with the small bowl and chicken meat into the pot. Cover the lid. Bring to the boil and turn to medium-low heat. Simmer for 4 hours. (Check the level of water from time to time. Add boiling water if needed)
4. Take the soup bowl out of the pot. Remove the small bowl and the chicken meat. Serve only the chicken juices in the soup bowl without adding any seasoning.

Remarks

If you don't want to use chicken, you can make pork essence the same way. Just use 300 g of lean pork instead of the chicken. After used to make this essence, the pork or chicken can be re-used in other soups for the rest of the family. Just add other ingredients such as vegetables and water.

This chicken essence cures Spleen-Asthenia, general weakness, dizziness, palpitation and menoxenia. Not only does it help recuperate after childbirth, it also serves as a great tonic for the elderly and those recovering from diseases. If you show signs of common cold or flu, do not consume chicken essence. Make pork essence instead.

Double-steamed common carp soup with red dates, black dates and black beans

Ingredients

1 common carp (with scales on)
5 red dates (de-seeded)
5 black dates
2 slices ginger
40 g black beans
boiling water

Seasoning

2 tbsp Shaoxing wine
salt to taste

Method

1. Rinse the common carp and wipe dry. Fry in a little oil until both sides golden. Rinse the fish in hot water to remove excessive oil.
2. Rinse both red dates and black dates. Fry the black beans in a hot dry wok until they crack open. Add cold water and bring to the boil. Cook for 3 minutes. Drain and rinse in cold water. Drain.
3. Put all ingredients into a double-steaming pot. Add boiling hot water.
4. Put the double-steaming pot into a steamer or a pot of boiling water. Double-steam for 4 hours. Season and serve.

Tips

This soup alleviates Asthenia, promotes diuresis, regulates blood chemistry and induces lactation.

Double-steamed silkie chicken soup with abalones and Bei Qi

Ingredients

4 live abalones
1 silkie chicken
30 g Bei Qi
30 g Huai Shan
10 g goji berries
5 g dried tangerine peel (soaked in water till soft)
5 dried longans (shelled and de-seeded)
2 slices ginger
boiling water

Seasoning

salt to taste

Method

1. Scrub the abalones well. Rinse and blanch in boiling water. Rinse in cold water. Shell them and remove the innards.
2. Dress and rinse the chicken. Remove the lungs. Blanch in boiling water. Rinse in cold water. Drain.
3. Put all ingredients into a double-steaming pot. Pour in boiling water to cover. Steam over medium heat for 4 hours. Serve.

Remarks

This soup nourishes the Yin, regulates blood chemistry, benefits the Spleen and promotes Qi (vital energy) flow.

Pork trotter soup with Tong Cao

Ingredients

1 pork trotter
10 g Tong Cao
10 sprigs spring onion
8 cups water

Seasoning

salt to taste

Method

1. Chop the pork trotter into pieces. Rinse well and blanch in boiling water. Rinse in cold water. Drain.
2. Rinse the Tong Cao. Rinse spring onion without removing its fibrous roots.
3. Put port trotter and Tong Cao into a pot. Add water. Bring to the boil over high heat. Turn to medium-low heat and boil for 2 hours. Put in spring onion with fibrous roots.
4. Turn to low heat and simmer for 10 minutes. Season with salt. Serve both the soup and the pork trotter.

Remarks

1. This is a therapeutic recipe for Qi- and Blood-Asthenia, no breast milk, light-coloured breast milk or insufficient breast milk supply after childbirth.
2. The white part of spring onion is rich in vitamin C. It warms the body, induces sweating and reduces fever. It is very effective in curing early symptoms of common cold and flu. The green leaves of spring onion contains ß-Carotene that maintains healthy mucus membranes and fights against respiratory infections. The pungent taste of spring onion comes from propylene sulfide, which helps eliminate toxins in the body, promotes detoxifying functions of the liver, boosts the immune system and has anti-oxidant functions.
3. Spices and aromatics helps induce sweat. The toxins in the body will also be excreted along with the sweat.

Veggie Papaya soup with peanuts and red dates

Ingredients

1 papaya
12 red dates (de-seeded)
6 dried figs (sliced)
80 g peanuts (with skins)
80 g cashew nuts
1 piece dried tangerine peel
10 cups water

Method

1. Peel and de-seed the papaya. Cut into chunks.
2. Put all ingredients into a pot. Add cold water. Boil for 2 hours. Serve.

Remarks

This soup is rich in protein, mineral, vitamins A, B and C. It smooths and moistens the skin, stops coughs, expels phlegm, improves eyesight, clears Heat, regulates digestive functions and promotes bowel movement.

 Chayote and pumpkin soup with tomato and shiitake mushrooms

Ingredients

2 chayotes
300 g pumpkin
2 tomatoes
60 g dried shiitake mushrooms
8 cups water

Seasoning

salt to taste

Method

1. Soak dried shiitake mushrooms in water until soft. Rinse well.
2. Peel and de-seed the chayotes and pumpkin. Cut into chunks.
3. Cut tomatoes into wedges with skin on.
4. Put all ingredients into a pot. Add water and bring to the boil over high heat. Turn to medium-low heat and simmer for 2 hours. Season with salt. Serve.

Remarks

This soup is nourishing and sweet in taste. It is a great source of vitamins.

Spinach soup with lotus seeds and white fungus

Ingredients

300 g spinach
100 g fresh lotus seeds
75 g white fungus
2 slices ginger
2 cups water

Seasoning

salt to taste

Method

1. Rinse the white fungus. Soak in water until soft. Rinse lotus seeds. Blanch in boiling water. Drain.
2. Rinse spinach. Keep the roots intact. Cut into short lengths.
3. Pour water into a pot. Add ginger, lotus seeds and white fungus. Cook for 10 minutes.
4. Put in spinach. Bring to the boil again. Season with salt. Serve.

Remarks

This soup replenishes the central Qi (vital energy), helps achieve calmness, and cures various ailments. The solid ingredients are highly nutritious and should be served along with the soup. Spinach is rich in iron, especially its roots. White fungus is rich in amino acids and collagen with polysaccharides. Lotus seeds is a healthful herb commonly used in health tonics, with abundant protein, lipids, carbohydrates, calcium, phosphorous and iron etc.

Steamed grouper slices with shiitake mushrooms and red dates

Ingredients

240 g grouper fillet
4 dried shiitake mushrooms (sliced)
4 red dates (de-seeded and sliced)
2 slices ginger (shredded)
1 sprig spring onion (shredded) or coriander (sectioned)

Marinade

1 tbsp ginger juice wine
1/2 tsp salt
ground white pepper
1 tsp caltrop starch
1 tbsp cooked oil

Method

1. Rinse the grouper fillet and slice it.
2. Add shiitake mushrooms, red dates and marinade to the grouper. Mix well and leave them for 15 minutes.
3. Arrange the grouper slices on a steaming plate. Transfer into a steamer. Steam over high heat for 6 to 8 minutes.
4. Drain any liquid on the plate. Sprinkle with shredded spring onion or sectioned coriander. Heat 2 tbsp of oil in a wok until hot. Drizzle over the grouper slices. Serve.

Stir-fried sliced pork with lily bulb and sugar snap peas

Ingredients

100 g lean pork
80 g fresh lily bulbs
80 g sugar snap peas
50 g re-hydrated cloud ear fungus
1 tsp ginger juice wine
1 clove garlic (sliced)

Marinade

1 tsp soy sauce
1/4 tsp caltrop starch
1 tsp cooked oil
1 tsp water

Seasoning

1/4 tsp salt
1/4 tsp sugar
1 tsp soy sauce
1 tsp caltrop starch
sesame oil
ground white pepper
5 tbsp water

Method

1. Rinse and slice pork. Add marinade and mix well. Set aside. Rinse the lily bulbs. Break them up into scales.
2. Blanch sugar snap peas and cloud ear fungus in boiling water. Rinse in cold water. Drain.
3. Heat 1 tbsp of oil in a wok. Stir-fry sliced pork until cooked through. Set aside. Rinse the wok. .
4. Heat 1 tbsp of oil in the same wok. Stir-fry garlic until fragrant. Put sugar snap peas and cloud ear fungus into the wok. Toss well. Add lily bulbs. Toss again. Put in the sliced pork and sizzle with ginger juice wine. Stir until the lily bulbs turn translucent. Add seasoning and toss well. Serve.

Rice wine chicken with sesame oil and ginger

Ingredients
1/2 pullet chicken
80 g old ginger
4 tbsp black sesame oil
1/2 cup glutinous rice wine or regular rice wine
1/2 cup water

Marinade
1 tsp salt
1 tsp cooked oil
1/2 tsp caltrop starch

Method
1. Rinse and dress the chicken. Chop into pieces. Slice the ginger thickly with skin on.
2. Heat a wok and add 3 tbsp of sesame oil. Stir-fry sliced ginger until lightly browned.
3. Put in the chicken. Stir to sear the chicken pieces on all sides. Add rice wine and water. Cook for 15 minutes. Serve. (It doesn't need any seasoning.)

Remarks
This recipe promotes blood circulation among women after childbirth. It also helps eliminate the toxins and wastes accumulated during the gestation period.

Fried cod fish in lemon honey sauce

Ingredients
2 cod fish fillets
2 shallots (sliced)
caltrop starch
wine

Marinade
1 tsp oil
1/2 tsp salt

Seasoning
2 tbsp soy sauce
1 tsp lemon juice
1 tbsp honey
2 tbsp water

Method
1. Rinse and wipe dry the cod fish. Add marinade and rub evenly. Leave it for 15 minutes. Coat the fish with caltrop starch lightly.
2. Heat a wok and put in 2 tbsp of oil. Fry the cod fish fillets on both sides until golden. Save on a serving plate.
3. Heat the same wok and add a little oil. Stir-fry shallot until fragrant. Sizzle with wine. Pour in the seasoning. Bring to the boil. Dribble the sauce over the cod fish. Serve.

Braised chicken in cinnamon honey soy sauce

Ingredients

1/2 Kamei chicken
 (local organic chicken, about 300 g)
10 g cinnamon
40 g ginger (sliced)
1/4 cup rice wine
8 red dates
(soaked in water till soft, de-seeded)

Marinade

1 tbsp ginger juice wine
1 tsp caltrop starch

Seasoning

2 tbsp honey
1/4 cup soy sauce
1/2 tsp salt
1/4 cup water

Method

1. Rinse the chicken and chop into pieces. Add marinade and mix well.
2. Heat a wok and add 2 tbsp of oil. Stir-fry ginger and cinnamon until fragrant. Put in the chicken. Sear on all sides until lightly browned. Sizzle with wine. Add seasoning and red dates. Bring to the boil and turn to low heat. Simmer for about 10 minutes until chicken is cooked through. Serve.

Soy marinated chicken

Ingredients

1 dressed chicken
2 shallots
2 slices ginger
2 whole star anise pods
600 g rock sugar
500 g light soy sauce
500 g dark soy sauce
1 tbsp Chinese rose wine

Method

1. Remove the lungs of the chicken. Rinse well. Blanch in boiling water and rinse in cold water. Drain well.
2. Rinse shallots, ginger and star anise. Put them into a deep pot. Add rock sugar and soy sauce. Cook over low heat until sugar dissolves. Pour in Chinese rose wine.
3. Cut off the chicken feet. Put the whole chicken into the soy marinade. Cook over medium-low heat for 10 minutes. Flip the chicken and cook for 10 more minutes.
4. Remove the cooked chicken from the marinade. Chop into pieces and serve. Strain the used soy marinade and leave it to cool completely. Pour it back into the bottle and use it as a seasoning.

N.B. The soy marinade bursts with flavours after the chicken is cooked in it. You may use it as a seasoning when making other dishes that call for soy sauce. It is so rich that you don't need to put in additional flavourings, such as chicken bouillon powder.

Pork trotters, ginger and hard-boiled eggs in sweet black vinegar

Ingredients

10 eggs
(or salted duck eggs)
2 pork trotters
(chopped into pieces)
1.8 kg ginger
1.2 kg unsweetened black vinegar
1.2 kg raw cane sugar slabs

Method

1. Put eggs into a pot of cold water with their shells on. Bring to the boil and keep cooking for 7 minutes. Rinse in cold water. Shell them.
2. Rinse the ginger and scrape off the skin with a metal spoon. Leave them to air-dry. Blanch pork trotters in boiling water. Drain and rinse in cold water.
3. Fry the ginger in a dry wok briefly.
4. In a clay pot or casserole, lay the ginger on the bottom. (Do NOT use a metal pot for this recipe.) Pour in vinegar. Bring to the boil. Turn to low heat and cook for 30 minutes. Add raw cane sugar slab. Cook until sugar dissolves. Keep on cooking over low heat for 30 minutes. Put in the pork trotters. Cook for 30 minutes. Turn off the heat and leave the lid on overnight. Reheat the next day over low heat for 30 minutes. Put in the shelled eggs. Boil for 10 minutes. Turn off the heat and cover the lid. Leave the eggs to soak in the vinegar until flavourful. Serve.

Remarks

Pork trotters are rich in protein, fat and carbohydrates. They promote metabolism, defy aging, stimulate breast milk production and beautify skin texture. Women may consume pork trotters in sweet black vinegar from the 12th day onward after delivery vaginally. Those who have a C-section should wait for 3 weeks before consuming. Those who adhere to a vegetarian diet may skip the pork trotters and eggs in the recipe.

Sweet and sour pork ribs

Ingredients

480 g pork belly ribs
2 slices ginger
2 shallots
1 whole star anise pod
1 piece cassia bark
1 tbsp ginger juice wine
2 cups water

Seasoning

3/4 tsp salt
1 tbsp soy sauce

Sauce

1 tsp Zhenjiang black vinegar
2 tbsp tomato paste
45 g raw cane sugar slab
(about 1/2 slab)

Method

1. Rinse and wipe dry the ribs. Cut into pieces about 2 inches long.
2. Heat 2 tbsp of oil in a wok. Fry the ribs until golden on both sides. Remove.
3. Add some oil in a wok. Put in ginger, shallot, star anise and cassia bark. Stir-fry until fragrant. Add the ribs. Sizzle with ginger juice wine. Add 2 cups of water and seasoning. Bring to the boil. Turn to low heat. Simmer for 40 minutes.
4. Stir in the sauce ingredients. Keep cooking over low heat until the sauce thickens. Transfer onto a serving plate.

Steamed fish in savoury egg custard

Ingredients

380 g fourfinger threadfin
(or any fish of your choice)
3 eggs
1 1/4 cups hot water
shredded spring onion

Marinade

1/2 tsp salt
1 tsp ginger juice wine

Seasoning

1 tsp salt

Method

1. Debone the fish. Rub marinade on the fish evenly. Leave it for 15 minutes.
2. Whisk the eggs. Add hot water to the seasoning.
3. Pour the hot water into the whisked egg while whisking continuously.
4. Put the fish into a deep steaming dish. Transfer into a steamer and steam over high heat for 5 minutes. Drain any liquid on the dish. Wipe dry.
5. Pour in the whisked egg mixture. Steam over medium-low heat for 5 minutes. Sprinkle with spring onion at the centre.
6. Heat some oil in a wok until hot. Drizzle over the spring onion. Serve.

Fish head soup with scrambled egg, ginger and rice wine

Ingredients

1 head of bighead carp
(cut into halves)
75 g black beans
20 g wood ear fungus
1 egg
160 g double-distilled rice wine
1/4 cup ginger juice
2 slices ginger (shredded)
4 cups water

Method

1. Fry the black beans in a dry wok until they crack open. Add 1 bowl of water. Boil briefly. Drain. Rinse in cold water. Drain again.
2. Soak wood ear fungus in water until soft. Shred coarsely. Whisk the egg and stir-fry in a little oil. Rinse the fish head and wipe dry. Fry in a little oil until golden. Set aside.
3. Heat a little oil in a pot. Stir-fry ginger and wood ear until fragrant. Add 4 cups of water. Put in ginger juice and wine. Finally, add fish head, black beans and scrambled egg.
4. Bring to the boil and turn to medium-low heat. Cook until the liquid reduces to 2 cups (for about 15 minutes). Serve.

Remarks

Bighead carp, as its name suggests, has a huge head that makes up about 2/5 of its body length. In between the flesh and the bones of its head, there is a blob of white gelatinous substance, which is rich in calcium, phosphorus, protein and fat. Consuming its head is believed to help cure headache and dizziness. Coupled with black beans and rice wine, this soup is good for those with weak Asthenic body and those lacking central Qi (vital energy).

Veggie ## Assorted vegetables on poached tofu

Ingredients

8 small dried shiitake mushrooms
80 g Shimeji mushrooms
80 g sugar snap peas
40 g carrot
40 g re-hydrated cloud ear fungus
80 g mung bean sprouts
2 slices ginger
1 cube tofu (about 400 g)

Seasoning

1 tbsp oyster sauce
1 tsp soy sauce
1 tsp vegetarian chicken bouillon powder
sesame oil
ground white pepper
1/2 tsp sugar
2 tsp caltrop starch
1/2 tsp water

Method

1. Soak shiitake mushrooms in water until soft. Drain and squeeze dry.
2. Rinse the sugar snap peas. Set aside. Cut carrot into floral-shaped slices or plain round slices.
3. Cut tofu into half-inch thick slices. Blanch in boiling water briefly. Drain and save on a serving dish.
4. Heat oil in a wok. Stir-fry ginger until fragrant. Put in shiitake mushrooms, shimeji mushrooms, sugar snap peas, carrot , cloud ear fungus and mung bean sprouts. Stir until cooked through. Sizzle with wine. Add seasoning and bring to the boil. Pour the mixture over the bed of tofu. Serve.

 Veggie ## Braised shiitake mushrooms with kelp

Ingredients

80 g dried kelp knots
40 g ginger (sliced)
160 g dried shiitake mushrooms
1 red chilli (shredded)
2 tbsp toasted sesames
5 g rock sugar
2 cups water

Seasoning

2 tbsp soy sauce
4 tbsp vegetarian oyster sauce
1 tbsp sesame oil

Method

1. Rinse the kelp knots. Soak in water until they swell (for about 2 hours). Drain and replace with fresh water from time to time.
2. Rinse the shiitake mushrooms. Soak in water until soft. Squeeze dry.
3. Heat 2 tbsp of oil in a wok. Stir-fry ginger and mushrooms until fragrant. Add 2 cups of water and 1 cube of rock sugar. Bring to the boil. Turn to low heat. Cook for 30 minutes.
4. Add seasoning and kelp knots from step 1. Cook for 15 minutes until the sauce reduces to 1/4 cup. Transfer onto a serving plate. Sprinkle with sesames and stir in shredded chilli. Serve hot or at room temperature.

Stir-fried mushroom duo with cashew nuts

Ingredients

80 g deep-fried cashew nuts
80 g fresh shiitake mushrooms
80 g straw mushrooms
120 g celery
6 slices carrot
(cut into floral shapes)
2 slices ginger
(cut into floral shapes)
2 coriander stems
(cut into short lengths)
1 tbsp ginger juice wine

Seasoning

1 tbsp vegetarian oyster sauce
1 tsp soy sauce
1/4 tsp sugar
1/4 tsp salt
sesame oil
ground white pepper
caltrop starch
4 tbsp water

Method

1. Rinse shiitake mushrooms. Cut off the stems. Blanch in boiling water briefly. Drain. Rinse in cold water. Drain again.
2. Cut celery into chunks. Make crisscross cuts on top of each straw mushroom. Blanch celery and straw mushrooms in boiling water separately. Rinse in cold water. Drain again.
3. Heat a wok and add 2 tbsp of oil. Stir-fry ginger until fragrant. Put in all ingredients. Toss quickly over high heat. Sizzle with ginger juice wine. Add seasoning and stir to coat well. Put in cashew nuts at last. Toss and serve.

 Stir-fried dried tofu with spinach and wood ear fungus

Ingredients

320 g spinach
10 g wood ear fungus
1 cube five-spice dried tofu
2 slices ginger (finely chopped)
1/4 cup water

Seasoning

1 tsp oyster sauce
1/4 tsp salt
1 tsp caltrop starch
2 tbsp water

Method

1. Rinse the spinach. Soak wood ear fungus in water until soft. Finely shred it. Slice the dried tofu.
2. Heat a wok and add 2 tbsp of oil. Stir-fry ginger, wood ear fungus and dried tofu. Add spinach and stir briefly. Pour in 1/4 cup of water (about 5 tbsp). Cover the lid and cook for 2 minutes. Set aside the ingredients. Drain any liquid in the wok.
3. Heat the wok again. Put all ingredients back in the wok. Add seasoning and cook until it thickens. Toss well and serve.

Veggie Braised pumpkin in soy and oyster sauce

Ingredients

300 g pumpkin
3 red dates (de-seeded)
1 tbsp diced ginger

Seasoning

1 tsp soy sauce
1 tsp oyster sauce
1/4 tsp salt
1/2 cup water

Method

1. Peel and dice pumpkin coarsely.
2. Rinse red dates. Slice them.
3. Heat 1 tbsp of oil in a wok. Stir-fry ginger and red dates until fragrant. Put in pumpkin. Stir well.
4. Add seasoning and stir again. Cover the lid and cook for 10 minutes until the sauce reduces. Serve.

 Veggie Stir-fried lotus root with Chinese celery

Ingredients

1 segment lotus root (about 300 g)
1 sprig Chinese celery
1 piece dried tofu
30 g red tarocurd
30 g ginger (shredded)
1 cup water

Seasoning

2 tsp soy sauce
1 tbsp oyster sauce
1 tsp sugar

Thickening glaze (mixed well)

1 tsp caltrop starch
2 tbsp water

Method

1. Peel and slice the lotus root. Rinse Chinese celery and cut into short lengths. Slice the dried tofu.
2. Heat 1 tbsp of oil in a wok. Stir-fry half of the shredded ginger until fragrant. Put in dried tofu. Toss well. Add Chinese celery. Toss and set aside.
3. In the same wok, heat 1 tbsp of oil. Stir-fry the remaining shredded ginger with the red tarocurd. Put in sliced lotus root. Toss well. Add water and seasoning. Cook until the sauce reduces to half.
4. Put the Chinese celery and dried tofu back in. Toss again. Stir in the thickening glaze. Mix well. Serve.

Veggie

Stir-fried pea sprouts with
shiitake mushrooms and oyster sauce

Ingredients

40 g dried shiitake mushrooms
300 g pea sprouts
1 tbsp shredded ginger
2 slices ginger
1/2 tsp rock sugar

Seasoning

1 tbsp vegetarian oyster sauce
1/2 cup stock
1 tsp soy sauce

Thickening glaze

1 1/2 tsp caltrop starch
4 tbsp water

Method

1. Rinse and soak shiitake mushrooms in water until soft. Squeeze dry and slice them.
2. Rinse the pea sprouts. Drain.
3. Heat 2 tbsp of oil in a wok. Stir-fry shredded ginger until fragrant. Put in pea sprouts and stir until they wilt and turn soft. Cover the lid and cook for 2 minutes. Strain and discard the juices and liquid in the wok. Transfer the pea sprouts on to a serving plate.
4. Heat 1 tbsp of oil in the same wok. Stir-fry the shiitake mushrooms until done and fragrant. Add seasoning and rock sugar and bring to the boil. Cook over low heat for 2 minutes. Stir in the thickening glaze. Mix well. Drizzle the pea sprouts with the glaze. Serve.

Remarks

Pea sprouts are rich in calcium, vitamins and carotene. From Chinese medical point of view, they promote diuresis, alleviate diarrhoea and ease oedema. They also aid digestion. Regular consumption helps keep the skin smooth and supple while adding a healthy glow to your skin.

 Toasted rice

Ingredients
400 g white rice

Method
Fry the rice in a clean dry wok over low heat until golden. Let cool. Save in an airtight bottle for later use.

Remarks
Toasted rice can be added to any herbal medicine, such as dried tangerine peel, red dates, Tong Cao or Bei Qi. Toasted rice tea is a great replacement of water. Just make sure you must make a fresh batch of toasted rice tea every day. Any leftover should be discarded the next day.

Veggie Toasted rice tea

Ingredients
12 cups water
3 tbsp toasted rice
1 piece dried tangerine peel

Method
1. Pour water into a pot. Add dried tangerine peel and 3 tbsp of toasted rice. (Do not rinse the toasted rice.)
2. Boil the mixture until the rice turns mushy. Save the tea in a thermal flask. Serve throughout the day when needed.

 Toasted rice tea with red dates

Ingredients

20 red dates (de-seeded)
2 slices ginger
2 tbsp toasted rice
12 cups water

Method

1. Rinse the red dates and put them into a pot. Add toasted rice and ginger.
2. Add water. Bring to the boil. Turn to low heat and cook for 1 hour. Serve.

Remarks

This tea expels Wind and warms the Stomach.

 Tong Cao tea with Bei Qi

Ingredients

10 g Tong Cao
15 g Bei Qi
2 tbsp toasted rice
12 cups water

Method

Rinse all ingredients. Put into a pot. Add water and bring to the boil. Turn to low heat and boil for 1 hour until the toasted rice turns mushy. Serve.

Sang Ji Sheng tea with hard-boiled eggs

Ingredients

120 g Sang Ji Sheng (mulberry mistletoe)
10 eggs (or to taste)
rock sugar
12 cups water

Method

1. Put eggs into a pot of cold water. Bring to the boil. Cook until the eggs are hard-boiled (for about 7 minutes). Soak in cold water. Shell and set aside.
2. Rinse Sang Ji Sheng. Put into a pot. Add water and bring to the boil. Turn to medium heat. Cook for 20 minutes. Put in the shelled eggs from step 1. Boil for 40 more minutes. Remove the Sang Ji Sheng with a strainer ladle. Discard.
3. Add rock sugar to the tea. Cook until sugar dissolves. Serve.

Remarks

From Chinese medical point of view, Sang Ji Sheng regulates blood chemistry, expels Wind, strengthens the Liver and the body, while benefitting the bones and sinews.

 # Creamy ground walnut sweet soup

Ingredients

2 cups shelled walnuts
8 red dates (de-seeded)
1/2 cup white rice
320 g rock sugar
8 cups water

Method

1. Rinse the walnuts. Soak in hot water until soft.
2. Cook red dates briefly in hot water. De-seed.
3. Soak white rice in water for at least 4 hours. Drain.
4. Put walnuts, red dates and white rice into a blender. Add 4 cups of water. Blend until fine. Strain through a fine mesh. Save the walnut and rice milk into a pot. (Do not use an iron pot for this recipe.)
5. Cook rock sugar in 4 cups of water until it dissolves. Pour in the strained walnut and rice milk from step 4. Turn on medium heat and keep stirring until it boils. Serve.

Double-steamed papaya sweet soup with milk

Ingredients

1 papaya
2 tbsp rock sugar (or to taste)
1 cup milk
1 cup water

Method

1. Rinse the papaya. Peel and de-seed. Cut into chunks. Put them into a double-steaming pot.
2. Put rock sugar and water into a pot. Boil until sugar dissolves. Pour this syrup into the double-steaming pot with the papaya.
3. Double-steam over high heat for 3 hours. Add milk and double-steam for 30 more minutes. Serve.

Remarks

1. Double-steamed papaya helps cure coughs, moistens the Lungs and beautifies the skin.
2. This sweet soup is suitable for all ages and everyone in the family. The recipe I put down here is enough to make 2 to 4 servings. Feel free to share it with your family.

Lotus root stuffed with glutinous rice in candied osmanthus syrup

Ingredients

200 g glutinous rice
1.28 kg lotus root
200 g white sugar
candied osmanthus

Method

1. Soak the glutinous rice in water overnight. Rinse and drain.
2. Rinse the lotus root. Cut off the last inch at the end of the lotus root.
3. Stuff the holes of the lotus root with the glutinous rice from step 1. Put on the last inch of the lotus root. Secure with toothpicks.
4. Put the lotus root into a pot. Add water to cover. Bring to the boil. Cook over medium-low heat for 2 hours.
5. Add white sugar and candied osmanthus. Cook for 1 hour until the syrup thickens. Set aside to let cool. Slice the lotus root and arrange on a serving plate.
6. Drizzle with the syrup from step 5. Serve.

Pumpkin sweet soup with coconut milk, sago and lily bulbs

Ingredients

300 g pumpkin
60 g sago pearls
160 g sugar
1 cup coconut milk
80 g fresh lily bulbs
6 cups water

Thickening glaze (mixed well)

2 tbsp corn starch
3 tbsp water

Method

1. Cut the pumpkin open. Remove the seeds. Steam until soft. Mash the flesh and set aside.
2. Rinse the fresh lily bulbs. Break them into scales. Set aside.
3. Boil sago pearls in water while stirring continuously. Cook until they turn transparent except a tiny white spot at the centre. Drain. Rinse in cold water to remove the excessive starch.
4. Boil 6 cups of water in a pot. Pour in the mashed pumpkin, sugar and sago pearls. Bring to the boil. Add thickening glaze while stirring continuously. Cook until it thickens. Add coconut milk and lily bulbs. Boil briefly. Serve.

Creamy ground cashew nut sweet soup

Ingredients

160 g cashew nuts
220 g rock sugar
5.3 cups water
30 g corn starch
120 g milk

Method

1. Rinse cashew nuts. Wipe dry. Bake in an oven until golden.
2. Put cashew nuts into a blender. Add 1 cup of water. Blend until fine. Pass the mixture through a fine mesh.
3. Save the cashew milk into a pot. Add water and bring to the boil. Add rock sugar and cook until it dissolves.
4. Mix corn starch with some water into a slurry. Stir into the sweet soup and cook until it thickens. Add milk at last. Serve.

Glutinous rice balls in candied osmanthus syrup with distillers grains

Ingredients

6 red dates
80 g glutinous rice flour
1 1/2 tbsp long-grain rice flour
120 g water
50 g distillers grains
1 cup rock sugar syrup
1 tsp candied osmanthus

Method

1. Mix glutinous and long-grain rice flour well. Add water and knead into smooth dough.
2. Put rock sugar syrup and red dates into a pot. Bring to the boil. Turn off the stove.
3. Roll the dough into small balls (about 1 cm in diameter). Blanch in boiling water until they float. Transfer the glutinous rice balls into the rock sugar syrup from step 2.
4. Add distillers grains and candied osmanthus into the syrup. Serve.

Almond milk with dried longans

Ingredients

120 g sweet almonds
40 g bitter almonds
100 g rice
240 g rock sugar (or to taste)
20 g dried longans (shelled and de-seeded)
8 cups water

Method

1. Rinse the almonds and rice. Soak them in water overnight. Drain.
2. Put in half or one-third of the water, almonds and rice into a blender at one time. Blend until fine. Strain and set aside the almond and rice milk.
3. Crush the rock sugar. Rinse the dried longans. Finely dice them. Put rock sugar, dried longans and the almond rice milk from step 2 into a pot. Cook over low heat and bring to the boil. Cook until sugar dissolves. Serve.

Double-steamed milk custard with ginger juice

Ingredients

1/2 cup water
2/3 cup sugar
2 tbsp ginger juice
160 g egg whites
250 g milk

Method

1. Boil the water and add sugar. Cook until it dissolves. Take 125 ml of the syrup. Add ginger juice and mix well.
2. Whisk the egg whites. Add milk and whisk until well incorporated.
3. Pour the hot syrup into the egg white mixture while whisking continuously. Pass the mixture through muslin cloth and divide among steaming bowls. Put them into a steamer and cover the lid. Steam over medium heat for 30 minutes. Serve.

Pork liver soup

Ingredients

150 g pork liver
1 slice ginger (shredded)
1 tbsp ginger juice
1 tbsp rice wine
1 cup water

Method

1. Rinse the pork liver. Finely chop it. Set aside.
2. Add ginger juice, rice wine and shredded ginger to the pork liver. Stir well.
3. Boil water. Stir in the pork liver. Boil until pork liver is cooked through. Strain and serve the soup.

Remarks

You may season with a pinch of salt.

Pork rib soup with white radish and lotus root

Ingredients

juice of 20 g white radish (with skin on)
1 segment lotus root (with skin on, with narrow nodes on both ends)
400 g pork ribs
8 bowls water

Seasoning

salt to taste

Method

1. Rinse the pork ribs and blanch in boiling water. Drain. Rinse the lotus root with skin on. Put pork ribs and lotus root into a pot.
2. Add white radish juice and water.
3. Bring to the boil. Turn to medium-low heat. Cook for 3 hours. Season with a pinch of salt. Serve.

Remarks

This soup prevents bloated stomach. You may serve this soup any time of the day in whatever serving size you prefer.

Roasted orange with salt

Ingredients

1 orange
salt
1 sheet aluminium foil

Method

1. Cut off the top of an orange to expose the pulp. Sprinkle with a pinch of salt.
2. Put the cut top on the orange again. Wrap the whole orange in aluminium foil.
3. Bake in an oven at the highest temperature for 20 minutes. Peel and serve the pulp while still hot.

Remarks

Roasted orange is effective in curing cold and flu, expelling phlegm, aiding digestion and promoting bowel movements.

Sheng Hua Soup (Postpartum herbal soup)

Promoting blood flow, easing Asthenia, aiding lochia discharge

According to ancient classics on herbal medicine, Sheng Hua Soup serves to promote blood flow. It is effective in replenishing the blood cells and aiding lochia discharge among women after childbirth. Sheng Hua Soup also activates antibodies and encourages uterine contractions.

Postpartum confinement is an invaluable folkloric wisdom and should be respected by all women. Childbirth is the biggest change a woman can go through in her life. You should seize this opportunity to recuperate and strengthen the body. You can live your life healthily and serenely thereafter.

Herbalist's advice

Sheng Hua Soup is a must-eat tonic for the novice moms right after the delivery of the baby. No matter you had a Cesarean section (C-section) or vaginal delivery, Sheng Hua Soup is meant to be consumed all the same. Follow these directions from herbalist:

- If you had a vaginal delivery, take Sheng Hua Soup on the first day after childbirth. Or, you may do it within the first week after childbirth. Just consume the soup once a day, for two consecutive days.

- If you had a C-section, wait till the fourth week after childbirth. Consume the soup once a day for two consecutive days.

Standard Ingredients

32 g Dang Gui Shen
60 g Dang Shen
16 g Huai Shan
8 g fried liquorice
12 g Chuan Xiong
12 g Bai Zhu
16 g fried Bai Shao
12 g Jiang Tan (charred ginger)
12 g peach kernels
12 g raw Di Huang
20 g cooked Di Huang
20 g Yun Ling
12 g Pu Huang (wrapped in muslin bag)
12 g Wu Ling Zhi

Method

Put all ingredients into a clay or ceramic pot. Add 5 1/2 cups of water. Boil over medium heat until 1 cup remains. Serve.

Points to note

- If a woman shows symptoms of common cold or flu, do not consume Sheng Hua Soup. Seek medical help from a doctor immediately.
- Try your best to avoid washing your hair in the first week after childbirth.
- If you have a sweet tooth, serve sweet soup with distillers grains.

** The invaluable information on this recipe is provided by Chinese herbalist Mr. Lo Sau Yu.

蔡潔儀，香港著名飪烹導師，擁有超過25 年教學經驗，教授學生數以萬計。著作種類繁多，計有《圖解點心製作教程》、《圖解中菜製作教程》、《香港經典小菜》等等，撰寫個人食譜超過 50 本，是全港飲食著作數目最多的作家，著作遍佈香港和海外。

蔡潔儀在飲食界的貢獻良多，旨在推廣飲食文化及培育有志於餐飲業的人士。曾任中華廚藝學院街頭美食班導師，現任僱員再培訓局專業培訓導師，並於 2012 年榮獲僱員再培訓局 ERB 優異導師獎。除此之外，她在業界被受推崇，於 2009 年榮獲國家級評委審議通過評為「中國國際名廚」，為第一位女性獲此殊榮；獲邀為廣東省佛山市餸朝餐飲服務有限公司高級顧問；在 2012 年榮獲第12 屆中國飯店金馬獎「中華英才五星勳章」、「中國餐飲新領軍人物」；2014 年榮獲香港飲食年鑑組織委員會「推動餐飲行業卓越大使」；2016 年榮獲香港飲食年鑑組織委員會「餐飲傳媒之星」；同年 11 月獲北京紅日時代文化傳播有限公司，授予為廚道匠心俱樂部創會港澳專家顧問；於 2017 年委任為澳門科技大學持續教育學院家政 / 廚藝專業文憑顧問。

作者簡介
蔡潔儀

坐月子調理良方
提升陪月員烹調技巧，
為產婦塑造健康體魄

Recuperative recipes for postpartum confinement

To train postnatal care helpers on their cooking skills for speedy recovery of the novice moms

作者
蔡潔儀

Author
Kitty Choi

顧問
盧壽如中醫師

Consultant
Lo Sau Yu

策劃/編輯

Project Editor
Catherine Tam

攝影

Photographer
Imagine Union

美術設計

Design
Ceci Chen / Charlotte Chau / Yu Cheung

出版者

Publisher

Forms Kitchen

香港鰂魚涌英皇道1065號
東達中心1305室

Room 1305, Eastern Centre, 1065 King's Road, Quarry Bay, Hong Kong.

電話　Tel: 2564 7511

傳真　Fax: 2565 5539

電郵　Email: info@wanlibk.com

網址　Web Site: http://www.wanlibk.com

http://www.facebook.com/wanlibk

瀏覽網站

會員申請

發行者

Distributor

香港聯合書刊物流有限公司
香港新界大埔汀麗路36號
中華商務印刷大廈3字樓

SUP Publishing Logistics (HK) Ltd .
3/F., C&C Building, 36 Ting Lai Road, Tai Po, N.T., Hong Kong

電話　Tel: 2150 2100

傳真　Fax: 2407 3062

電郵　Email: info@suplogistics.com.hk

承印者

Printer

百樂門印刷有限公司

Paramount Printing Company Limited

出版日期

Publishing Date

二零一七年五月第一次印刷

First print in May 2017